Ronan O'Dowd © 2011
(100 pages)

Ronan O'Dowd's Designer Books in Photonic Engineering:

Photonics Handbook 1

"Photonics Handbook: Part 1 Broadband Fibres"

"Photonics Handbook: Part 1 Broadband Fibres"

Author

Ronan O'Dowd PhD SMIEEE is Professor Emeritus Photonic Engineering at UCD Dublin, Ireland where he taught and researched Optoelectronics and Photonics for three decades until 2010. He has several breakthrough papers in topics such as tunable semiconductor lasers and optical communications, including the millennium 2001 paper proving a dense comb of 2000 wavelength channels could be transmitted in a single fibre using the same semiconductor laser (ref *IEEE Jnl.S.T. Quantum Electronics Mar 2001)*). Many of his students have proceeded to successful careers in academia and the photonics industry worldwide.

By the same author:

Physics Science of Action   Gill and Macmillan 1984

Tips to use this guidebook.

This series of books for photonic system designers will cover the subsystems that make up an optical communications link. These are the transmitter, fibre channel and receiver and since the fibre design sets the specifications and key criteria to be implemented at either end we tackle that first in this part 1 of the book.

The guide is formatted with *You Do* exercises and answers are provided. These are intended to be part of the learning process that will take the student to Engineering degree high level over what may be about a 30 hour degree module where 6 additional hours are set aside for practical work. The hardware link kit from OptoSci, where optical fibre dispersion etc can be measured over fibre reels, is very useful for laboratory explorations.

The later part 2 covering laser, detector and system design for broadband, can constitute a further full module.

"Photonics Handbook: Part 1 Broadband Fibres"

CONTENTS

     Practical work may use the OptoSci kit which contains transmitter, fibre reel and receiver along with test equipment for dispersion, bit-error rate etc. Section 7 recommends sample experiments.

Notes:

You Do exercises should be attempted especially by the self-taught student using this guide and regardless of your confidence in your answer quality. Having then read the answer provided you should again attempt it.

Diagrams are simple line-style that the student should re-draw.

There are sample examination questions at the end using standard mathematical relations.

1   Design Targets

The greatest obstacle to broadband is dispersion. Data bits spread in time and overlap causing errors.

Radio communication uses frequencies of hundreds of kilohertz to megahertz in the form of electromagnetic waves that carry information by modulation. Modulation impresses information bits onto the waves.

Mobile and satellite systems use higher frequency microwaves at 1-10 GHz

At this x1000 times higher frequency the information carrying bandwidth or bw is in turn x1000 times greater, a fact of the science.

Infra-red light is at over 100 THz so we expect $10^{14}/10^{10}$ or bw capability over 10,000 wider again.

So what must we do to actually achieve this BROADBAND capability using light instead of microwaves?

Firstly we choose fibre optics over metal for many reasons.
List of advantages of optical fibre over metal:

[1]   Low loss (0.2 dB / km); much greater span.
[2]   High bandwidth (many GHz.km); greater bit-rate.
[3]   Interference free unlike metal as glass is an insulator.
[4]   Low cross-talk as glass does not leak to nearby fibres.
[5]   Light weight (kg versus tonnes); easier installation in ducts and vehicles (e.g. aircraft / automobiles).
[6]   Small size (0.125 mm so many fibres per cable); suits existing ducts and cable pulling.

[7] Spark free; suits industrial environments.
[8] Glass raw material is silicate or silica, SiO_2, and is abundant unlike metals.

You Do...For you to do now
Ex 1: Draw a long glass rod and show light rays entering and then travelling down it by total internal reflection. Show faster and slower light paths.
Is Snell's Law of refraction relevant anywhere in your figure?

Answer to Ex 1

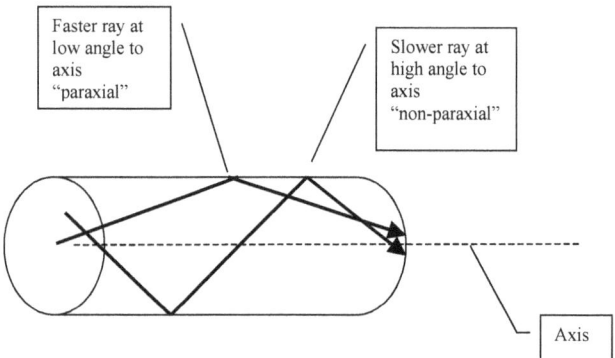

Faster ray at low angle to axis "paraxial"

Slower ray at high angle to axis "non-paraxial"

Axis

Figure 1A.

Snell's Law of refraction is relevant where the rays enter and exit the rod.
It also applies to the rays that reflect by total internal reflection TIR since these must strike the boundary at

greater than the critical angle for refraction which is calculated for a ray that refracted out of the rod at 90 degrees.

You Do...For you to do next

Ex 2: Draw the slowest and fastest rays in the rod.
Use the glass refractive index of $n_1 = 1.5$ to estimate the path difference per metre length of rod where the diameter is 0.1 mm

Answer to Ex 2

The five figures below help visualise the ray that glances the boundary at just greater than critical angle C. Medium 1 is glass (index 1.5 or n_1) and medium 2 is air (index 1.0 or n_0).
The second figure shows how to calculate C using Snell's law

$$\sin 90 \deg / \sin C = n_1 / n_0$$
$$\rightarrow \quad C = \sin^{-1} n_0/n_1 = \sin^{-1}(1/1.5) = 42 \deg$$

The third figure traces that ray back to before it entered the rod and uses Snell again to calculate A the angle within which rays are accepted and later guided by TIR. Beyond that capture zone light may enter but is then lost from the guide by refraction. The glancing ray in the middle figure is at (90 – C) deg to the axis so that for entrance facet refraction

$$n_1 / n_0 = \sin A / \sin (90-C)$$
$$\sin A = 1.5 \sin (90-42) = 1.115$$

This result is greater than 1 and therefore called non-physical as it cannot happen. This is because the angle 48 deg exceeds the critical angle.

Rays at that divergence from the axis would never have entered in the first place. A lesson here is that the maths contains warning signs if we are alert to them.

The fourth figure shows that angle A actually defines in three dimensions or 3D the acceptance cone of light that can enter and be guided by TIR. The same cone applies by symmetry at the exit facet. Since later we will add a cladding glass around the core of the fibre whose index is only 1% lower than the main glass core the angle A is quite small in practice and hence transmitters and receivers must be extremely precisely aligned with optical fibres to avoid large coupling losses. That is a challenge for mechanical engineers. If the cladding had index $n_2 = 1.49$ and we repeat the above calculation we find $C = \sin^{-1} (1.49/1.5) = 83.4$ deg so that $90-C = 6.6$ deg and that is well within the range of 42 deg for the air-glass entrance facet. In that case the third diagram shows that $\sin A/\sin 6.6 = 1.5/1$ giving $A = \sin^{-1} (1.5\sin 6.6) = 13.3$ deg and that result is now a fairly low aperture. Angle C within the guide is very large now (as a quick calculation by you using Snell can show) so all TIR rays are paraxial (meaning close to the axis) when a cladding glass is included in the design. We will therefore plan to create a glass cladding later as we proceed with improvements.

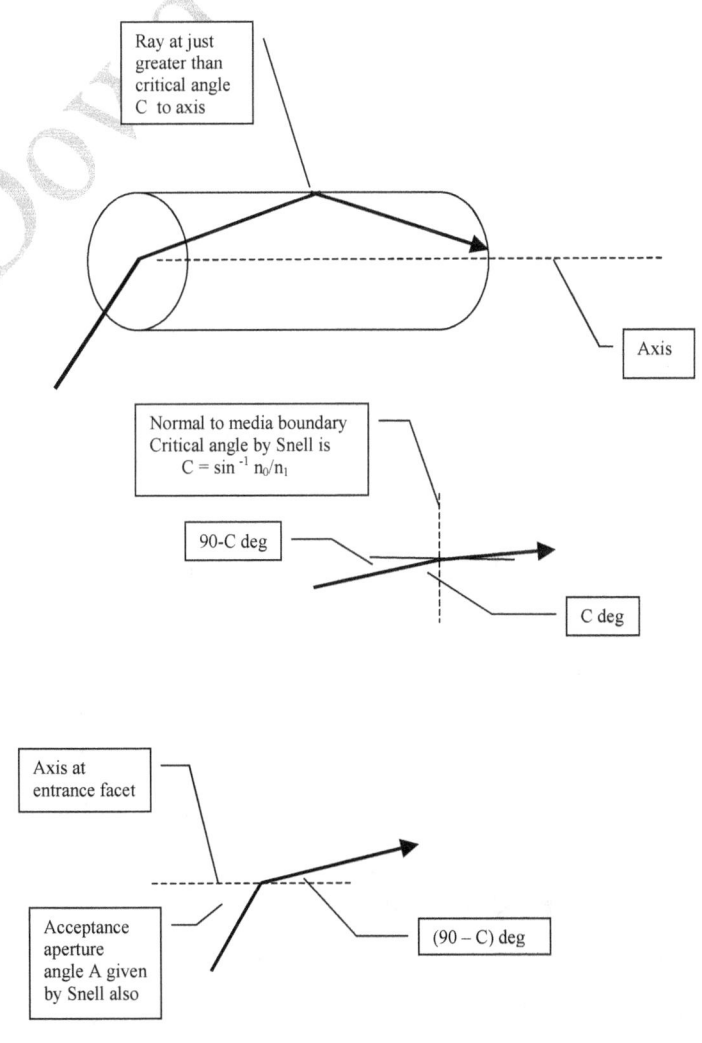

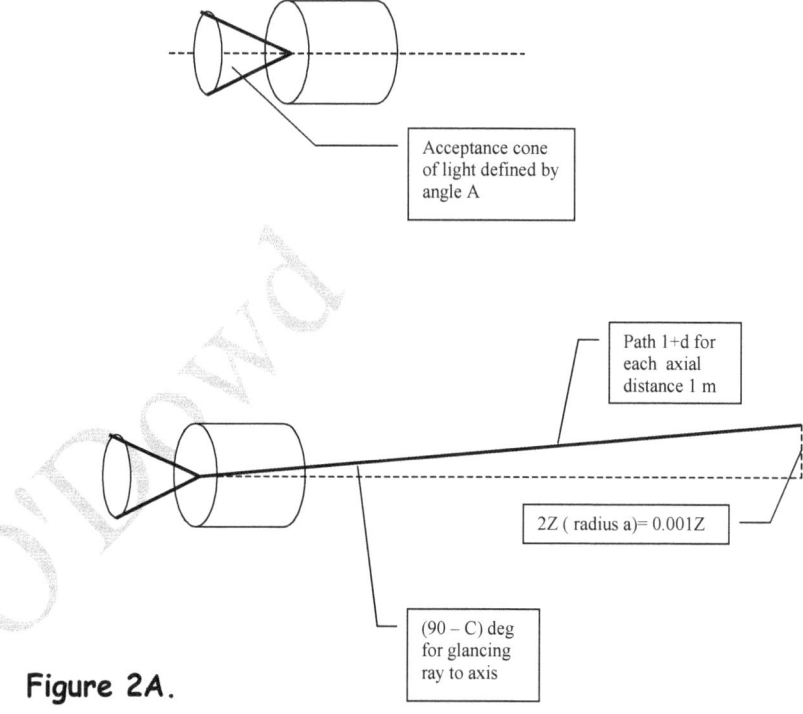

Acceptance cone
of light defined by
angle A

Path 1+d for
each axial
distance 1 m

2Z (radius a)= 0.001Z

(90 – C) deg
for glancing
ray to axis

Figure 2A.

The fifth figure shows a glancing ray path 1+d where d is the extra distance in metres travelled for each 1 m down the axis. Note the units there: metres per metre. The effective dispersion per metre of fibre, d, is calculated from

$$\cos(90-C) = 1/(1+d) \text{ or } \cos^{-1}[1/(1+d)] = 90-C$$

Using C for the core-cladding from above in this equation gives dispersion d, the path difference in metres per metre travelled down the axis.

Finally if we divide by speed of light c/n_1 the dispersion is in seconds for each metre down the axis.

For curiosity we may also find from this triangle the number Z of zig-zags or reflections for the ray per metre:

$$(1+d)^2 = 1^2 + (2aZ)^2$$

Note the $2aZ$ since for each reflection the ray then travels $2a$ across the guide to reach the boundary again. We can transpose each zig-zag segment of the path into a straight line for computation without altering the maths. This linear transformation can easily be shown using graph paper.

Low Attenuation Glass

A rod of conventional glass loses light so fast that after a few metres it is dark. Even a window pane which is that thick is impenetrable to light, visible or infra-red, IR.

The second important requirement after bandwidth is near lossless glass to achieve several km range. Ideally for inter-city communications we need > 100 km with minimum loss.

If the receiver (a well designed RX) can detect 100 times less light than is launched by the laser transmitter or TX then 99% less power is tolerable. The remaining 1% is 1/10 x 1/10 times so in decibels (ten times log base 10) that is minus 10 dB and then minus another 10 dB down or -20 dB total. This is because log 10 of 1/10 is -1; times 10 gives -10 dB.

You do...
Ex 3: Power on dB scale.
A semiconductor laser emits 2 mW of light power from each end of the chip. Translate to the commonly used photonic engineering decibel or dB units.

Answer to Ex 3
We take either 1 mW or 1 μW as reference power. For the former the laser emits 2 mW which is ×2 times the reference and the $\log_{10}$ of that is 0.3 called 0.3 bel. There are 10 decibel to one bel so the power is 10×0.3 giving 3 decibels relative to a milliwatt. This is written 3 dBm where the m refers to a milliwatt, mW.

Alternatively relative to one microwatt or a 1 μW reference power the laser emits ×2000 times and $\log_{10}$ of that is 3.3 giving 33 dBμ where the μ refers to a microwatt, μW.

Each additional 10 dB added means ×10 times greater power in comparison to the reference so 30 dB added means 10×10×10 or ×1000 times greater relative to the reference power.

1 dBm equals 30 dBμ and 3 dBm is the same as 33 dBμ. Remember that multiplication becomes addition in terms of logarithms. That is why we use the dB scale in the first place.

Since $\log_{10} 2 = 0.3$ or 3dB then doubling a power adds 3 dB.
The laser above emits at each chip end equally so for total power out of it add 3 dB giving 3+3=6 dBm or 33+3=36 dBμ in total.

Laser transmitters at **1.5 μm wavelength**, λ, are well developed using light emitting diodes made of compound semiconductors and not of silicon but rather semiconductor quaternary alloys of indium In, gallium Ga, arsenic As and phosphorous P.
Since frequency $\nu = c/\lambda$ we compute $3\times10^{8}/1.5\times10^{-6}$ giving a light frequency ν of 2×10^{14} Hz so this laser is at 200 THz or 200,000 GHz
So the capacity is 200,000 times broader than for a microwave system at 1 GHz. That is certainly broadband at the TX. A high speed response at the receiver RX is also needed so the photodiode and its amplifier circuit must be carefully designed too.

We also require range of several km.

The product of both these specifications, frequency and distance, in **GHz.km** (gigahertz kilometre) is therefore a measure of both important requirements in our target design.

Equally, the inverse **ns/km** (nanoseconds per kilometre) can be used but in that case the smaller the better.

You do...

Ex 4: Think about the units ns/km and figure out what they actually mean in the physical system.

Answer to Ex 4

As data travels down the fibre the existence of fast and slower pathways causes dispersion d calculated above. This is measured in nanoseconds or picoseconds spreading of the light pulses and the further it travels the greater the spread or dispersion. The dispersion after each kilometre is measured in nanoseconds per kilometre, ns / km. That is the inverse of the bandwidth times distance product in GHz.km.

○

Power Loss

Let us tackle the range problem first from the two targets, distance and bandwidth..

The attenuation or loss in glass is due to many factors that classify into absorption and scattering.

Attenuation is high at UV and shorter visible wavelengths due to Rayleigh scattering and slowly drops off according to 4th power of λ as we enter infra-red or IR until we reach 1.5 μm. After this it rises again by absorption as we enter the longer IR.

That is why the TX laser needs to be at *1.5 μm minimum loss.*

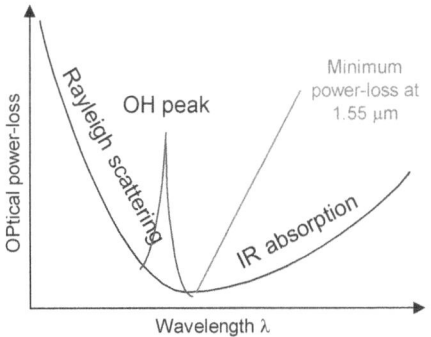

Figure 3A. Master power-loss curve for silicate glass SiO$_2$.

Rayleigh scattering, Figure 3A, falls with $1/\lambda^4$ as we move to longer wavelength from blue while after the red colours IR absorption takes over beyond 1.55 μm wavelength. There is a localised peak at 1.4 μm due to absorption by unavoidable, tiny, residual water content (OH-bond absorption); that peak separates the 1.3 μm or S-band (for short) from the 1.5 μm or C-band (for conventional). Compound semiconductor alloys InGaAsP of different elemental mix are needed to transmit at these different wavelengths and that science is called band-gap engineering. The optimum low-loss region around 1.55 μm is called the C-band for conventional and it can accommodate a comb of many closely spaced optical frequencies or wavelengths called dense wavelength division multiplexing DWDM. A single optical fibre may carry 50 DWDM channels or more in the C-band.

Chemical Vapour Deposition CVD

Fibre manufacture involves extremely pure gas ingredients unlike window panes that are made from treated sand. The gases carry elements Silicon and Oxygen that react in a flame to form pure SiO_2 or silicon dioxide which is glass. Other carrier gases bring dopants to the reaction like boron or phosphorous to alter the refractive index according to precisely programmed design. In that way we can create an INDEX PROFILE across the rod diameter of the so-called preform that results from the depositing solids (Figure 4A). The index profile formed by precise vapour deposition determines whether the fibre will be step-index multimode, graded-index multimode or monomode (single mode).

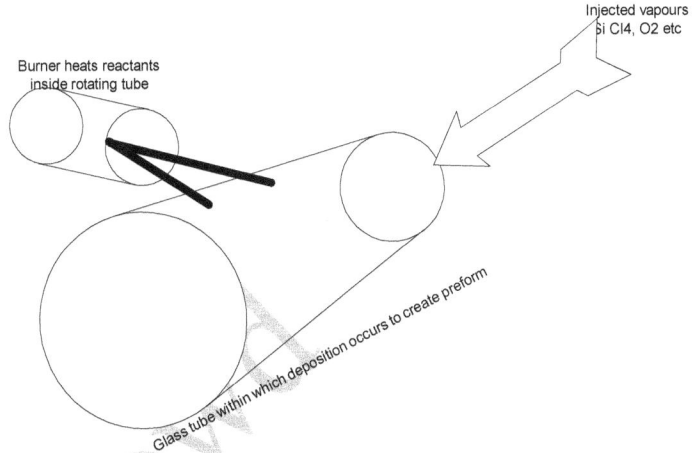

Figure 4A. Glass perform grown by chemical vapour deposition CVD.

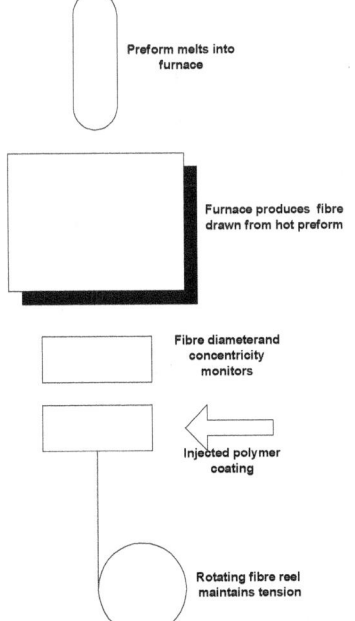

Figure 5A. Drawing tower to produce fibre reel of 1 or more kilometres from glass preform.

To draw the bulky preform into thin fibre a drawing tower is used (Figure5A). The cylindrical sample of about 1m long by several cm diameter grown by chemical vapour deposition CVD can be heated and drawn as a 0.1 mm strand onto a reel carrying several km length of fibre. This fibre has diameter 125 or 100 μm and its refractive index profile mimics that of the original bulk preform. Next the fibre is cabled for strength and protection from impurity ingress or mechanical impacts. Loose-tube cables have internal plastic grooves in an extruded central form that let the fibre sit without stress. Several fibres may be accommodated in each cable.

To join two reels of fibre a precision splicer is required as the 100 μm thick ends must match within <1 μm axial alignment accuracy (Figure6A).
When a re-usable joint is needed a precision mechanical connector that is demountable and of similar accurate capability is essential (Figure7A).

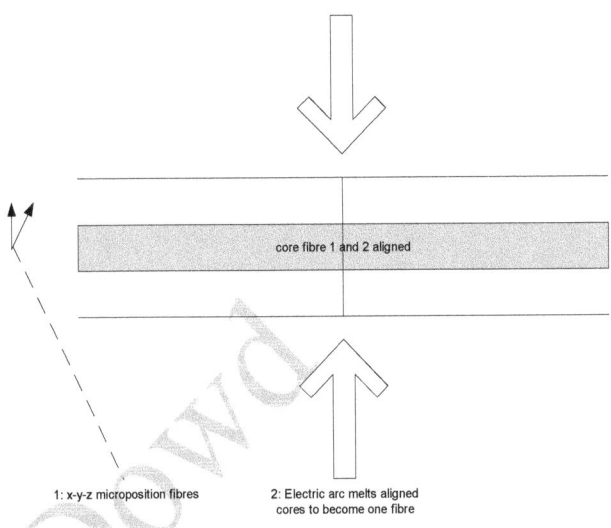

core fibre 1 and 2 aligned

1: x-y-z microposition fibres

2: Electric arc melts aligned cores to become one fibre

Figure6A. Fusion splicer for fibres; coupling loss 0.1 dB.

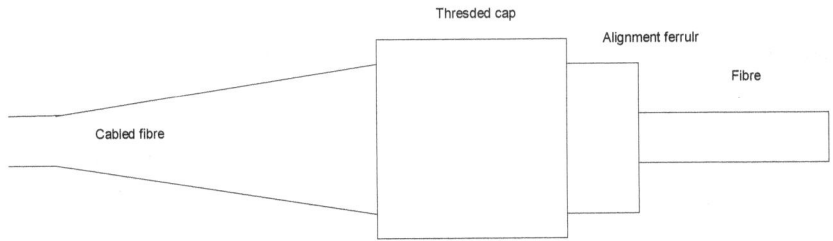

Thresded cap

Alignment ferrulr

Fibre

Cabled fibre

Figure 7A. Precision connector for de-mountable joints.

The reel or installed cable can be tested by optical time-domain reflectometry, OTDR, where a fast laser pulse of high power is launched and the back-scattered and reflected light is measured to identify cracks as well as the gradual fall-off in power with increasing range (Figure8A).

The underlying gradual fall-off is a logarithmic process measured in decibels per kilometre dB / km while reflection losses at couplers are in dB.

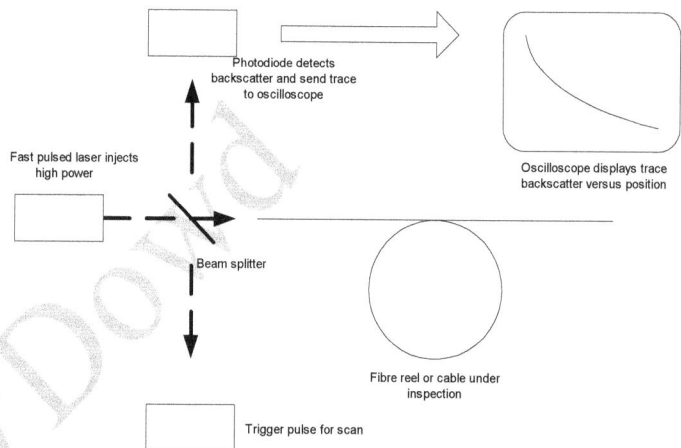

Photodiode detects backscatter and send trace to oscilloscope

Fast pulsed laser injects high power

Oscilloscope displays trace backscatter versus position

Beam splitter

Fibre reel or cable under inspection

Trigger pulse for scan

Figure 8A.   Optical time domain reflectometer OTDR.

You Do...

Show that a perform of 1 m long and 1 cm diameter can be drawn onto a reel of 100 μm diameter fibre with length 10 km.

You do...

Ex 5: Link design example with power-loss budget measured in dBm

A laser launches 4 dBm into a fibre with 0.4 dB/km loss at that wavelength. The span of 10 km reaches a detector with 3 dBμ sensitivity but with a 1 dB coupling loss. There are two connectors each with a 1.5 dB coupling loss. Allow for temperature and time degradation to assess final excess power if any.

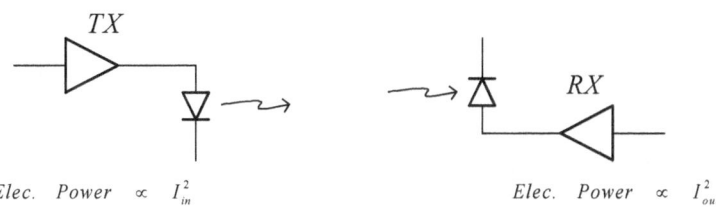

Figure 9A. Transmitter TX and receiver RX in fibre link.

At the fibre ends the TX has a laser diode emitting light while the RX has a photodiode receiving light. Electrical power depends on square of current but light power is proportional to current itself as electrons produce photons.

Answer to Ex 5

Launched at TX laser: +4 dBm

Reciever RX sensitivity: 3 dBμ or -27 dBm
Margin: 31 dB

Fibre loss for 10 km: 0.4x10= 4 dB
Connectors loss: 1.5x2= 3dB
Splice loss: no splices= 0 dB
Detector coupling loss: 1 dB
Allowance for temperature: 3 dB or factor 2 fall off
Allowance for time degradation 3 dB
Total attenuation: 14 dB
Excess power: 31-14 = 17 dB

Conclusion: each 3 dB means double the required power so there is vast power capacity here. A cheaper, less sensitive detector could be selected or, since sensitivity falls off with data rate, a much higher data speed could be used to improve broadband capacity as more customers join this part of the network or as they demand higher quality service.

o

Summary 1

The fibre we deploy must be of the highest calibre glass created by CVD in a fashion similar to silicon chips with the difference that we need silicon dioxide, called silicate or silica. While silicon or Si is a semiconductor the glass is an insulator, SiO_2. This extremely pure glass has exceptionally low attenuation, about 0.2 dB/km permitting several tens of km range without any repeater or booster. In that case a link of 100 km would have a loss of 0.2x100 = 20 dB and so the RX input power would be 100 times lower than that launched at the TX laser. A 3 mW laser could therefore deliver 30 μW at the detector in this example.

2   Rays and LP Modes

Rays give the direction in which the wavefront of linearly polarised light travels. This is called an *LP mode*.

Communications systems have three primary sub-systems. These are transmitter TX, fibre channel and receiver RX.
Each has its own response time and the inverse of that time is the sub-system bandwidth limit bw. The response times combine like Gaussian responses to give a lower full bw as follows:
System response time = [TX time ^{2} + channel time ^{2} + RX time ^{2}]$^{1/2}$
The overall response time as given here is called the root-mean-square, rms. The inverse is the system bandwidth bw.

So TWO BUDGET ANALYSES are needed in system design, power-loss and bandwidth.

You Do...

Ex 6: Link design example 2, bandwidth budget.

A link will use non-return to zero or NRZ data encoding where the system rise time is 0.7/data rate. It is a telephone cable at 140 Mbit/s with a 0.5 ns laser response. The total dispersion in the fibre span is 1.22 ns as measured for example with a laboratory OptoSci kit. The PIN detector has a 1 ns response. Perform a rise-time or bw budget.

Answer to Ex 6

Data rate:	140 Mbit/s
Required system rise time:	
for NRZ 0.7 / 0.14×10^9 =	5 ns
Laser rise time	0.5 ns
Fibre dispersion:	1.22 ns
Detector rise time:	1.0 ns
Sum of squares:	2.74
System root-mean-square rise time:	rms 1.66 ns

Conclusion: the response speed is fast and more than sufficient.

o

A POWER BUDGET is required to select the fibre for the system. This fibre must also satisfy the BANDWIDTH BUDGET so it is designed by engineers who grow the glass somewhat like silicon chips are made. The BROADBAND link will need fibre of excellent REFRACTIVE INDEX PROFILE so that the fastest rays of light and the slowest rays keep in pace and do not spread data time-wise. Alternatively we may let the fast rays only prevail with single mode operation.

What refractive index profile inside a tiny optical fibre will deliver broadband? The only sure way to answer that is to model the rays with mathematics. But a clue to what we expect the analysis to deliver (*and if it does not we will suspect our mathematics*) can be reasoned out as follows:

The dispersion d in ns/m or ns/km that we estimated for the primitive design example at Ex 2 occurred quite simply because the slow and fast rays or linearly polarised LP modes have different path lengths in the glass. Recall that the rod was 1 m x 0.1 mm. Let us call the diameter D = 0.1 mm and radius a will be half of that.

You Do...

Ex 7 :Reducing modal dispersion.

Re-draw the rod and rays again labelling fast and slow LP modes.

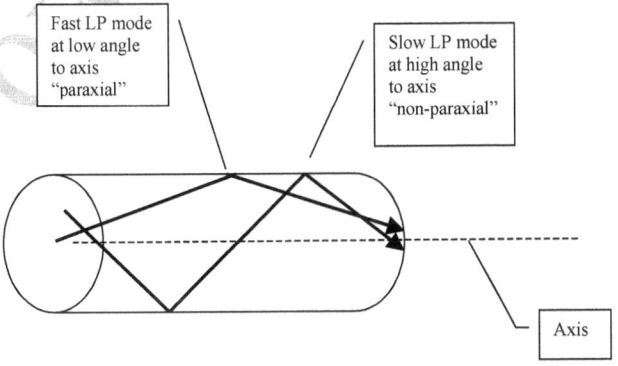

Figure 10A.   Fast and slow modes of the guide.

What property of glass could now speed up only the slow modes?

Answer to Ex 7

Speed of light in glass v = c/n where c is fixed at $3x10^8$ so that refractive index n must be made lower to speed up slower modes. Where do these slow modes reside more often but fast paraxial modes do not?

The answer is near edges of the glass rod. Therefore we shall reduce n at and towards the edges of the guide, perhaps gradually like this:

OLD Step index   NEW Graded index

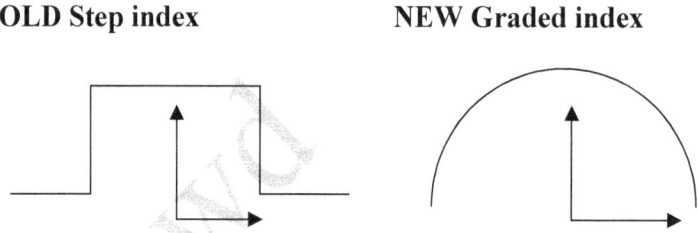

Figure 11A. Refractive index profiles; horizontal arrows are radial position r from 0 to a, vertical arrows represent increasing index n(r) as a function of radial distance from the axis.

So a graded index GI instead of a step index SI profile should reduce the spread of mode speeds called the MODAL DISPERSION and the mathematical analysis should predict that kind of result.

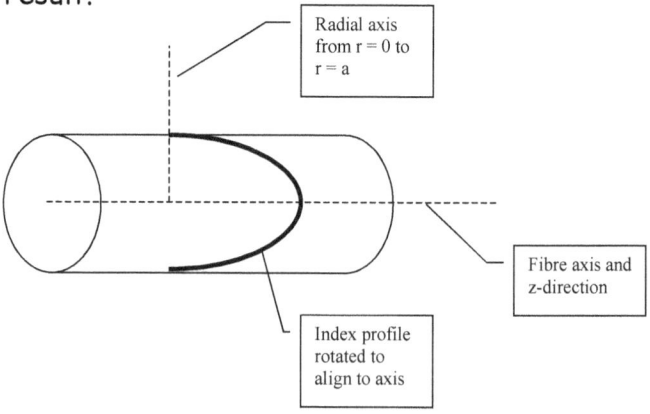

Radial axis from r = 0 to r = a

Fibre axis and z-direction

Index profile rotated to align to axis

Figure 12A. Graded index plot from last figure turned on its side and superimposed to line up with the fibre optic guide axis.

Another Improvement

Core-cladding guide: To protect the glass core itself from moisture and contamination a coating of polymer is laid during the drawing tower process (Figure 5A). But this protective polymer layer absorbs light so a second pure glass outer region, the cladding, is grown around the core during the CVD process. Then the polymer is coated on by the fibre drawing tower. This can avoid polymer absorption loss. We must therefore extend the profile by as much again, it turns out, and the index plot becomes:

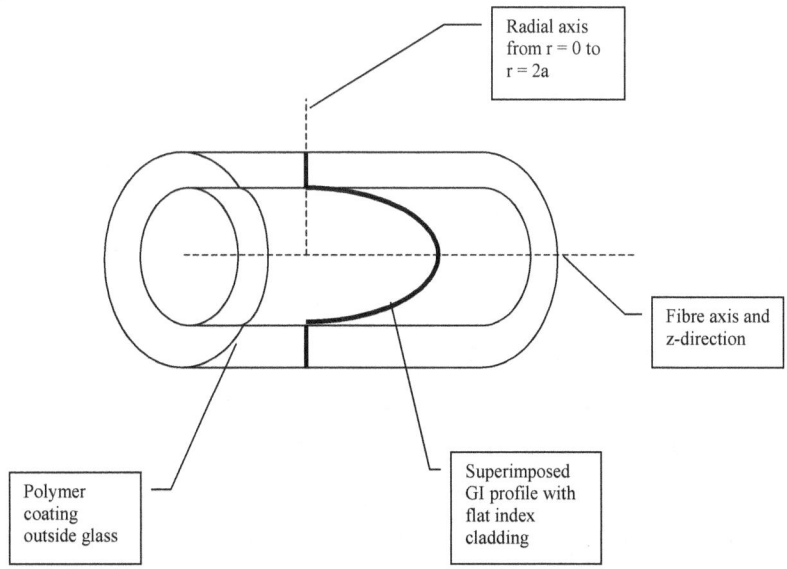

Radial axis from r = 0 to r = 2a

Fibre axis and z-direction

Polymer coating outside glass

Superimposed GI profile with flat index cladding

Figure 13A. Graded index fibre with flat index cladding and polymer coating.

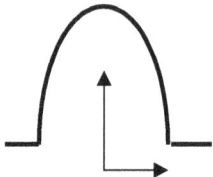

Figure 14A. The new profile has a GI core surrounded by a flat index cladding. Vertical arrow shows rising index and horizontal arrow is radial r from 0 to a.

You Do...

Ex 8: Re-read Ex 2. Calculate dispersion where n_1 or n_{core} = 1.5; n_2 or $n_{cladding}$ = 1.49; n_0 or n_{air} = 1 and where the slowest ray enters the fibre from air at angle A deg to the axis.

Use Snell's law to find angle inside fibre for slowest LP mode and also the new critical angle for TIR.

What is the dispersion d?

Draw the cone of light defined by A that can enter the fibre from air using the above calculations by tracing a TIR ray backwards from inside to air outside the fibre end facet.

In practice the end facets are cleaved with a special tool to a shiny finish to minimise end coupling losses.

Answer to Ex 8.

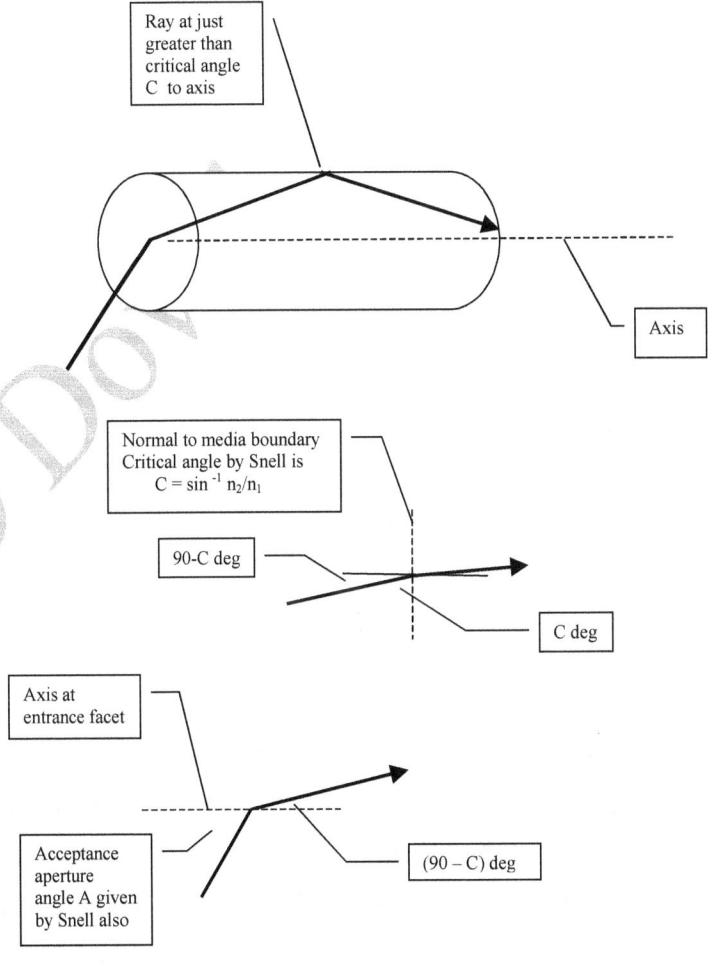

Ray at just greater than critical angle C to axis

Axis

Normal to media boundary
Critical angle by Snell is
$C = \sin^{-1} n_2/n_1$

90-C deg

C deg

Axis at entrance facet

Acceptance aperture angle A given by Snell also

$(90 - C)$ deg

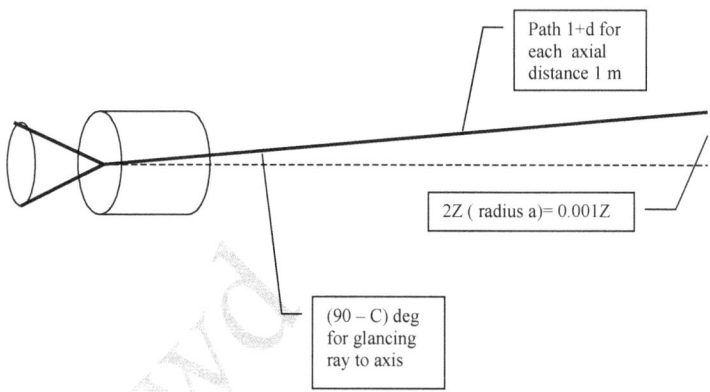

Figure 15A Calculating "primitive" dispersion by ray method.

Re-read Ex 2 first.

Recalculate: $C = \sin^{-1}(n_2/n_1 = \sin^{-1}(1.49/1.5) = 83.4$ deg

$$90 - C = 6.6 \text{ deg}$$

By Snell at the entrance facet

$$A = 13.3 \text{ deg}$$

$$\cos(90-C) = 1/(1+d) \text{ or } \cos^{-1}[1/(1+d)] = 90-C$$

[Or simply $\sin C = 1/(1+d) = n_2/n_1$ etc]

Now $\cos 6.6 = 0.9933$ so $1+d = 1/0.9933 = 1.0067$ metre per metre. This gives d=0.0067 metre per metre

Alternatively $d = 0.0067/(c/n_1) = 34$ ps/m or 34 ns/km

Bandwidth distance product is the inverse: 0.03 GHz.km

Finally $(1+d)^2 = 1^2 + (2aZ)^2$ and $2a = 50 \times 10$ -6 therefore:

$$Z = (0.52 \times 10^6)^{1/2} = 748$$

This result is large because the diameter is so small.

○

Summary 2

The fibre should be chemically grown SiO_2 glass to achieve high purity. The refractive index profile should have a higher index at the axis gradually falling towards the edge and with a lower, flat index cladding around the core. This design reduces modal dispersion considerably improving bw.

A cladding of glass at a flat index about 1 % below the axial value contributes and separates the lit-up core from the polymer coating that protects the extremely pure glass from light absorbing contaminants to sustain loss budget over many years. The cladding has a great effect on the critical angle. In turn the acceptance angle is reduced but as a result many higher modes are not guided. This in conjunction with the reduced index provided by a graded profile means that further from the axis the dispersion is greatly reduced. That is GI multimode fibre and is an improvement on SI multimode profile in terms of bandwidth by a factor of x 10^5

Ray and Wave Optics

Ray optics was used to reach this point but we need to move to WAVE OPTICS now to get a design engineer's perspective. A ray gives the direction in which an electromagnetic WAVEFRONT moves. A plane wavefront therefore has parallel rays. A spherical wavefront has radial rays emanating from a point. The WAVE EQUATION from Maxwell's theory of electromagnetism describes the motion through space (x,y.z) in time (t).

That equation uses a vector operator in space so let us briefly review a few fundamentals. From electromagnetic theory we know that electric field and voltage are related:

$$E_r = - \delta V/dr$$

When r is a vector in xyz space and i,j,k are unit vectors in these three directions this becomes:

$$E = iE_x + jE_y + kE_z$$
$$= - (i\delta/\delta x + j\delta/\delta y + k\delta/\delta z)V$$

The operator in brackets is called grad for gradient of the voltage or also del and symbolised as Δ and it is a vector since i, j and k are unit vectors in x, y and z directions:

$$\Delta V = gradV = (i\delta/\delta x + j\delta/\delta y + k\delta/\delta z)V$$

An inverted delta can be used if you prefer.

The dot product of Δ with itself is Δ^2:

$$\Delta^2 = \Delta.\Delta = \delta^2/\delta x^2 + \delta^2/\delta y^2 + \delta^2/\delta z2$$

This is shown by simply multiplying out the bracketed terms $(i\delta/\delta x + j\delta/\delta y + k\delta/\delta z)$ with itself to produce six terms and noting that three terms are zero being cross products of orthogonal unit vectors like ixj etc and that $i^2 = j^2 = k^2 = 1$.

The common form of Maxwells equation for an E field in the vertical or y-direction is:

$$\Delta^2 E_y = \mu_0\varepsilon_0 \; \delta^2 E_y/\delta^2 t^2$$
$$= (\delta^2/\delta x^2 + \delta^2/\delta y^2 + \delta^2/\delta z^2) \; E_y$$

The Δ^2 operator may be written in other co-ordinate systems such as spherical but the one we will opt for is cylindrical to match the shape of optical fibres.

In cylindrical polar co-ordinates r, ϕ, z we find:

$$\Delta^2 = \delta^2/\delta r^2 + 1/r \, . \, \delta/\delta r + 1/r^2 \, . \, \delta^2/\delta\phi^2 + \delta^2/\delta z^2$$

Another expression we will deploy is the speed of light in a medium like glass. The speed of light c in vacuum or air is related to the progress of electrical and magnetic effects. In a dielectric it is the electrical permittivity ε and magnetic permeability μ of the medium that determine the speed v:

$$1/c^2 = \mu_0\varepsilon_0$$

In glass where r means "relative to vacuum" this becomes:

$$1/v^2 = \mu_0\mu_r\varepsilon_0\varepsilon_r$$

For transparent materials $\mu_r = 1$ (approx.) so dividing these two equations we get:

$$v/c = (1/\varepsilon_r)^{1/2}$$

This ratio of speeds in the medium c/v is also its refractive index. Therefore relative permittivity ε_r is related to refractive index n:

$$n = (\varepsilon_r)^{1/2}$$

A further expression that we need to account for power loss when light enters the glass from air was established by Fresnel. The reflected and transmitted portions are related to refractive index by:

$$R = [(n-1)/(n+1)]^2 \quad \text{and } T = 1\text{-}R$$

You Do...
Ex 9: Draw plane and spherical wavefronts and the associated directional rays.
Answwer to Ex 9

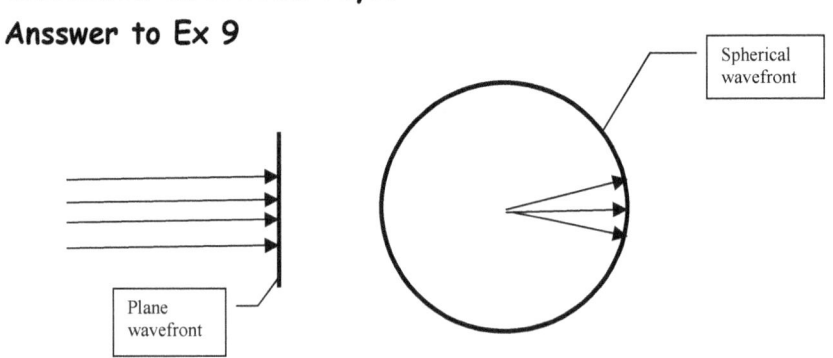

Figure 16A. Wavefronts and rays.

○

3   Mathematical Model

WKB (Wentzel Kramers Brillouin) Method

This technique was applied in 1973 to optical waveguides by Gloge and Marcatili at Bell Labs. We require the dispersion or bandwidth or impulse response for the cylindrical optical waveguide. Start with the wave equation where we assume index variation is negligible over a wavelength distance.

You Do...

Ex 10: Show index variation is negligible over a wavelength distance to be the case using the following data: radius fibre 25 μm, core index 1.51, cladding index 1.5 and laser wavelength 1550 nm.

Ans Ex 10 at end handbook.

$$\nabla^2 \psi = n^2 \varepsilon_0 \mu_0 \frac{\delta^2 \psi}{\delta t^2} \qquad (1)$$

This wave equation 1 relates the variation in space, del-squared, of the optical field psi or ψ (left hand side, lhs) to its variation in time (right hand side, rhs). The light will be contained inside the core of a cylindrical fibre so that the operator del-squared on the lhs will be more convenient in cylindrical polar co-ordinates giving equation 2.

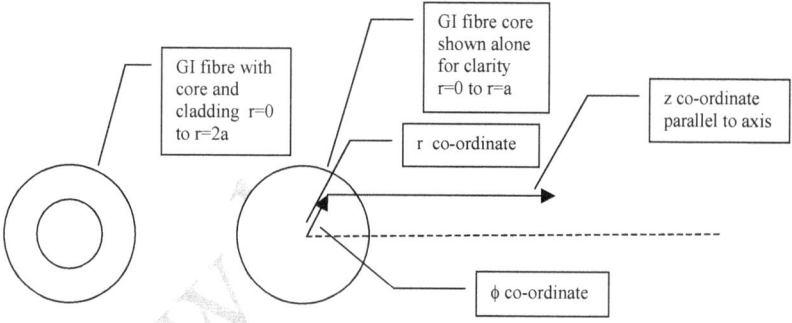

Figure 1. The fibre has core and cladding, left, but for brevity and clarity we will show the core alone as we proceed. Cylindrical polar co-ordinates in the fibre are r,φ,z. Move out radially r inside the core of radius a, then swing around polar or azimuthal by φ, and travel down parallel to the axis distance z.

$$\Rightarrow \frac{\delta^2\psi}{\delta r^2} + \frac{1}{r}\frac{\delta\psi}{\delta r} + \frac{1}{r^2}\frac{\delta^2\psi}{\delta\phi^2} + \frac{\delta^2\psi}{\delta z^2}$$
$$= n^2\varepsilon_0\mu_0\frac{\delta^2\psi}{\delta t^2}$$

(2)

Now we have the variation in space on the lhs in terms of radial r, polar or azimuthal φ and finally axial position z or distance down the fibre from the launch point. The rays that indicate direction of travel bounce back and forth by TIR at the core interface with the cladding. Outward and returned waves interfere as is the nature of light. Since there is cylindrical symmetry, as the light progresses we should expect the patterns of bright and dark created by the waves as they interfere to have cylindrical symmetry themselves.

Postulate solution: $\psi = F(r)\cos(\nu\,\phi)e^{i(\omega t - \beta z)}$ **MODE**

This has the expected cylindrical symmetry in the radial r and azimuthal or ϕ senses where the integer ν describes the polar or azimuthal mode pattern or its periodicity. There are ν pattern repeats as we swing round the circle from 0 to 2π radians so this is the **azimuthal mode number**.

The cosine in respect of this polar position means the brightness rises and falls periodically in a circle while the pattern also varies radially. Such patterns also appear for example on the skin of a vibrating drum. We insert this form of field, the mode, into the wave equation next.

$$\Rightarrow \quad \frac{d^2F}{dr^2} + \frac{1}{r}\frac{dF}{dr} + \left(n^2 k^2 - \beta^2 - \frac{\nu^2}{r^2}\right)F = 0 \qquad (3)$$

You Do...

Ex 11: Differentiate the postulated solution, MODE, once and then twice with respect to each of the co-ordinates r, ϕ and z. Insert into the wave equation 2 and tidy up to get (3).

Ans Ex11 at end handbook.

Hint:

$$\psi = F(r)\cos(v\,\phi)e^{i(\omega t - \beta z)}$$

$$\frac{\delta\psi}{\delta r} =$$

$$\frac{\delta^2\psi}{\delta r^2} =$$

$$\frac{\delta^2\psi}{\delta\phi^2} =$$

$$\frac{\delta^2\psi}{\delta z^2} =$$

$$\Rightarrow \frac{d^2F}{dr^2} + \frac{1}{r}\frac{dF}{dr} + \left(n^2k^2 - \beta^2 - \frac{v^2}{r^2}\right)F = 0$$

○

Both cylindrical and spherical symmetry are common in engineering and science. In the WKB method it was found that solving this equation (3) in linear x-space is tedious but a transformation to exponentially or log compressed space greatly eases the mathematics. Next transform to a log type of space:

$$r = ae^x \quad \text{and hence} \quad dr = ae^x dx$$

$$\Rightarrow \frac{d^2F}{dx^2} + \left(\kappa^2 a^2 e^{2x} - v^2\right)F = 0 \tag{4}$$

Where $\qquad \kappa^2 = n^2k^2 - \beta^2$

You do...

Ex 12: Use partial differentiation to get from Eq (3) to Eq(4)
Ans Ex12 at end handbook.

○

Observe the form of this equation (4); it suggests some kind of oscillatory solution. This is because it resembles the second order differential equation for simple harmonic motion:

$$\frac{d^2F}{dx^2} + (constsnt)F = 0$$

Therefore we give the solution amplitude and phase properties.

Use $F(x) = A(x)e^{iS(x)}$ for amplitude A and phase S of the field. Take second derivative of this product (with respect to) wrt x and insert into (4) to get (5):

$$\Rightarrow \quad \frac{d^2A}{dx^2} + 2i\frac{dA}{dx}\frac{dS}{dx} - A\left(\frac{dS}{dx}\right)^2 + iA\frac{d^2S}{dx^2} + \left(x^2a^2e^{2x} - v^2\right)A = 0 \qquad (5)$$

Break into real and imaginary parts and tidy up to get:

$$\Rightarrow \quad \begin{cases} \dfrac{dS}{dx} = \left(\kappa^2a^2e^{2x} - v^2\right)^{\frac{1}{2}} & (6) \\[2mm] \dfrac{d}{dx}\left[A^2\dfrac{dS}{dx}\right] = 0 \quad \Rightarrow \quad A = C\left[\dfrac{dS}{dx}\right]^{-\frac{1}{2}} & (7) \end{cases}$$

$$\Rightarrow \quad A(x) = \frac{C}{\left(\kappa^2a^2e^{2x} - v^2\right)^{\frac{1}{4}}} \qquad (8)$$

$$(6) \Rightarrow S(x) = \int_{x_1}^x \left(\kappa^2a^2e^{2x} - v^2\right)^{\frac{1}{2}} dx \qquad (9)$$

Return to "linear" r-space from log type of space using:

$$r = ae^x \quad \text{and hence} \quad dr = ae^x dx$$

This gives: $\qquad A(r) = \dfrac{C}{\left[\left(n^2k^2 - \beta^2\right)r^2 - v^2\right]^{\frac{1}{4}}} \qquad (10)$

41

and $$S(r) = \int_{r_1} \left[(n^2 k^2 - \beta^2) - \frac{v^2}{r^2} \right]^{\frac{1}{2}} dr \qquad (11)$$

The phase S in $F(x) = A(x) e^{is(x)}$ should be real to produce an oscillatory solution, otherwise we get an exponentially decaying result called an evanescent wave as a result of $i^2 = -1$. For a sustained field or guided wave the expression in equation (11) must therefore be real; this passes to imaginary when the square root term in parenthesis passes through zero and goes negative. The transition from guided mode occurs at:

"Turning points"...... $$\left[n^2(r) k^2 - \beta^2 \right] - \frac{v^2}{r^2} = 0 \qquad (12)$$

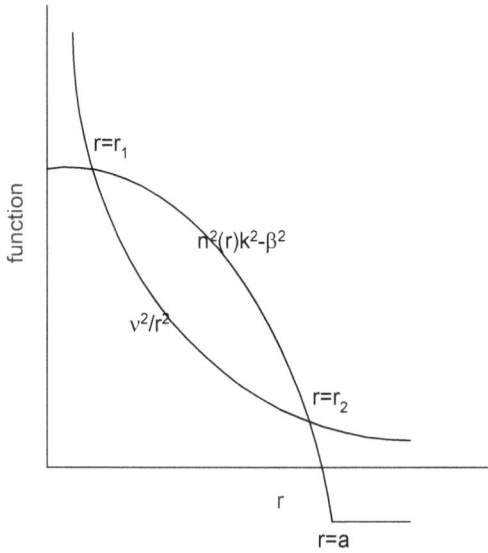

Figure 2. Plots to identify turning points.

To examine this we can draw plots for $[n^2(r)k^2 - \beta^2]$ and $\frac{v^2}{r^2}$ in order to identify where the first is above the second plot (Figure 2) so that the difference is positive. Reading inside the square brackets the first plot resembles the index profile squared times a constant and then pull it down by β-squared. The second plot is a constant over r-squared. This Figure 2 produces two values r_1 and r_2 defining an annular ring within which there is a real phase and hence a guided mode or ray, where the latter is the direction of travel for that mode (Figure 3).

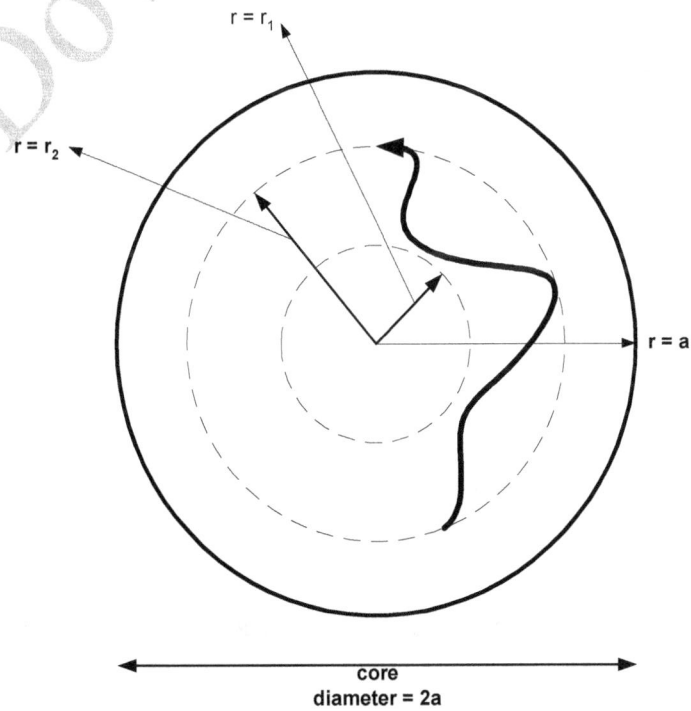

Figure 3. Azimuthal path of a guided ray within core section. The ray is spiralling down the fibre z-axis always contained inside the turning points for that mode r_1 and r_2.

In Figure 3 we see how each different mode will have different turning points as given by solution to equation (12). This means there is a dark spot at the centre for some modes, whenever r_1 is non-zero. There is also a dark ring outside r_2 if it is less than a. This is actually an interference pattern created by all the reflected waves that propagate down the core by diffraction.

Using equations (10) and (11) in the postulated solution for equation (2) namely Mode $\psi = F(r)\cos(v\,\phi)e^{i(\omega t-\beta z)}$ along with $F(r)=A(r)e^{is(r)}$ gives the result for the guided field:

$$\psi = \frac{1}{2}A(r)\{\exp[i(\omega t-\beta z-v\phi+S(r))]+\exp[i(\omega t-\beta z+v\phi+S(r))]\} \qquad (13)$$

You Do...
Ex 13: Produce equation (13) from (2) using (11).
Ans Ex 13 at end handbook.

The full standing wave should include by addition the negative square root from equation (11) and hence:

$$\psi = A(r)\cos(v\phi)\cos[S(r)]e^{i(\omega t-\beta z)} \qquad (14)$$

44

You do...

Ex 14: Produce Eq(14) from Eq(Mode) and (13)

Ans Ex 14 at end handbook.

o

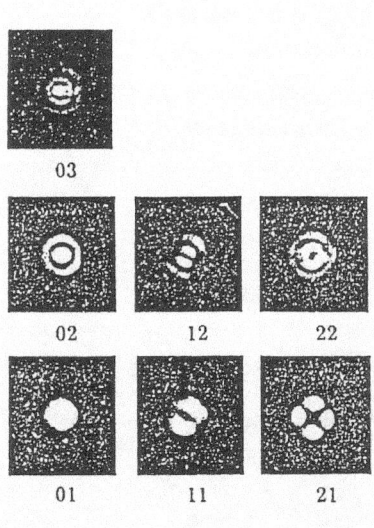

Figure 4   Linearly polarised LP modes $LP_{\nu\mu}$ (As measured by Stolen in 1976 after optically isolating each mode). The fundamental mode LP_{01} is at bottom left.   Study carefully and relate each to the mode integers.

The permitted patterns in Figure 4 are indicated each by two integers that are subscripts for that LP mode. When a ray traverses a full radial path from r_1 to r_2 and back the phase should return through an integer number μ of 2π, this being the second or **radial mode number**; additionally there are two 90 degree reflections at the turning points totalling another π. Hence we may write the phase condition:

$$2\int_{r_1}^{r_2}\left[n^2(r)k^2 - \beta^2 - \frac{v^2}{r^2}\right]^{\frac{1}{2}} dr = (2\mu + 1)\pi \qquad (15)$$

$\Rightarrow$ WKB *eigenvalue equation* including the second radial integer for the mode μ becomes:

$$\Rightarrow \qquad \int_{r_1}^{r_2}\left[n^2(r)k^2 - \beta^2 - \frac{v^2}{r^2}\right]^{\frac{1}{2}} dr = \left(\mu + \frac{1}{2}\right)\pi \qquad (16)$$

The two integers v and μ that describe a selected mode pattern are inserted into Eq (16) and the solution provides the propagation constant β. This process can be repeated for other index profiles n(r) or for other wavelengths contained in k.

SUMMARY 3

LP modes are the solutions to the wave equation in the cylindrical guide and these allowed light patterns have a set of maxima and minima or bright and dark spots. Each mode pattern $LP_{\nu\mu}$ has cylindrical symmetry like the acoustic patterns on the skin of a cylindrical drum. Each LP mode has two integers that describe its pattern, the azimuthal ν and the radial μ giving maxima totals simply ν and $2\mu+1$ and in polar and radial directions respectively.

4   Dispersion Mechanisms

The WKB eigenvalue equation is so called because we insert fibre profile, laser wavelength, selected mode numbers and solve for propagation constant β of that mode. Then the distribution of β-values gives the modal dispersion i.e. arrival time in nanoseconds of fastest versus slowest modes after each km length of guide. Knowing the dispersion allows us to calculate the bandwidth.

Recall: Dispersion (ns/km) = Inverse bandwidth x distance product (1/GHz.km)

Modal Dispersion

At Bell Laboratories in the last century much research was done to produce the design for the first viable optical fibres satisfying commercial loss and bandwidth requirements. We still use the Gloge-Marcatili approach and their expression for a general index profile fibre is:

$$
\left.
\begin{aligned}
n(r) &= \left[1 - 2\Delta\left(\frac{r}{a}\right)^{\gamma}\right]^{\frac{1}{2}} \quad \cdots\cdots \quad r < a \\
&= n_1 \left[1 - 2\Delta\right]^{\frac{1}{2}} \quad \cdots\cdots \quad r \geq a
\end{aligned}
\right\} \tag{17}
$$

Here $\Delta = \left(n_1^2 - n_2^2\right)/2n_1^2$ which measures the relative drop in refractive index from core to cladding interface or the difference squared across the core profile.

This Δ approximates to $(n_1 - n_2)/n_1$ and is therefore below 1%.

The index is assumed to fall from n_1 at $r = 0$ to n_2 at $r = a$ and γ (power exponent) is **profile shape-factor** or rate of fall-off. This was found by the researchers to be optimised at

$$\gamma_{opt} \approx 2(1 - 1.2\Delta) \qquad (18)$$

Beyond the core the second part of equation (17) holds and shows the profile is flat at the interface value, also given by the first equation when $r = a$. (Check this yourself by putting $r = a$ in the first equation).

Since Δ is very small (try typical values for n_1 and n_2) Eq(18) yields a shape factor $\gamma = 2$ for optimum graded index profile multimode fibre. When we insert $\gamma = 2$ into Eq(17) the result is the typical expression for a parabola so a **parabolic index profile** is the optimum shape for multimode fibre.

For a single λ source like a good quality laser this was then shown to give the dispersion for graded index (G.I) fibre

$$\left[\Delta\tau_{opt} \approx \frac{Ln_1}{8c}\Delta^2\right] \qquad \text{G.I. modal dispersion} \qquad (19)$$

What this Eq(19) represents is the difference in time on the lhs between travel by the fastest and slowest modes over a length L of fibre.

You Do...

Ex 15: Insert typical values as used previously into Eq(19) to get dispersion in parabolic index guide over each km of fibre.

Ans Ex 15 at end handbook.

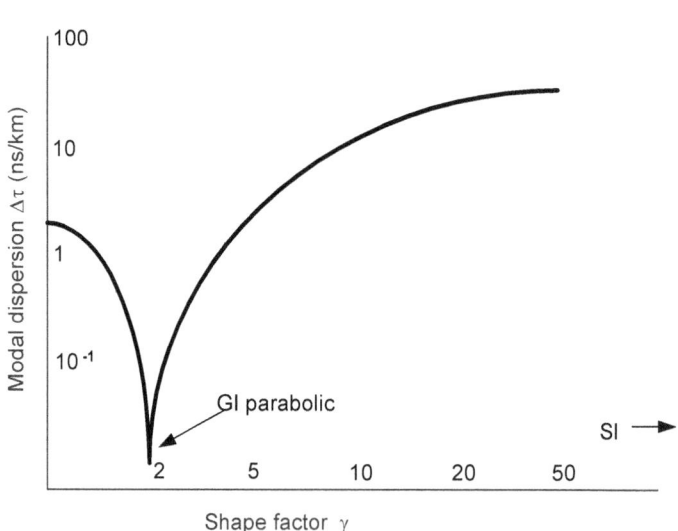

Figure 5   Modal   dispersion plot versus index profile shape factor. Observe the sharp cusp at GI parabolic represents improvement of 10^4 over SI.

This fibre dispersion due to modes may be re-calculated for other profiles by using a range of shape factors and the results are plotted in Figure 5. The γ-value is the ordinate and ranges from very low through $\gamma = 2$, GI parabolic, all the way to $\gamma = $ infinity for step-index SI as provided by Eq(17).This last point is because for any $r < a$ the fraction r/a to the power of infinity is zero so the *index is flat* at n_1 inside the core and then steps down abruptly to n_2 at the interface where r= a.

You Do...Ex: Show this easily using Eq(17) for core and cladding.

The plot is spread over a very wide range of resultant dispersions, so much in fact that a vertical log axis is used. There is a very sharp cusp at $\gamma = 2$ where the dispersion in ns/km is minimum. At the far right where γ approaches infinity, or step-index fibre, the dispersion reaches an ultimate value. Between the optimised parabolic GI profile and the SI case the dispersion improves by a whopping 100,000 so that bandwidth for GI parabolic is 10^5 times greater when we learn to grow this optimised profile. The challenge there following the maths is computer control of the precision valves that gradually open/close during glass perform growth.

Total Dispersion

There are further contributions to the spreading of the light pulses in the guide.

Total fibre dispersion comprises:

Modal dispersion as above

Profile dispersion where the profile is optimised, say parabolic, at one wavelength only but looks a different shape, not fully optimised, at others because index is wavelength dependent.

Wave-guide dispersion where dimensions of the guide play a role somewhat as happens when microwaves travel down a metal conduit.

Material dispersion, that for silica glass is *zero at 1300 nm* as was discovered by measuring the refractive index of glass over many colours. It was found that chromatic dispersion reverses (blue faster than red instead of red faster than blue, see Figure 6) at a specific wavelength depending on the material. This chromatic effect can overwhelm modal dispersion, particularly in GI fibre where the modal effect is greatly reduced or in monomode fibre where it is eliminated.

Chromatic dispersion is the combination of material along with the smaller profile effect above since both are colour dependent.

In classical optics of the nineteenth century anomalous dispersion was investigated and the plots of n versus λ recorded. The plot can be differentiated to yield data for material dispersion Δτ:

$$\left[\Delta\tau = \frac{L\lambda}{c}\left(\frac{d^2n}{d\lambda^2}\right)\Delta\lambda \right] \qquad \text{Material dispersion} \quad (20)$$

The difference in time of travel for longer versus shorter wavelengths was found to depend on the second rate of change of index with wavelength as shown in Eq(20). It also increased with the spread of wavelengths in the light source $\Delta\lambda$ and obviously with the physical length of the fibre link L or more specifically with L/c which is the time to travel L at speed c. Furthermore, as we move to longer wavelength λ the dispersion degrades. Eq (20) summarises all of this.

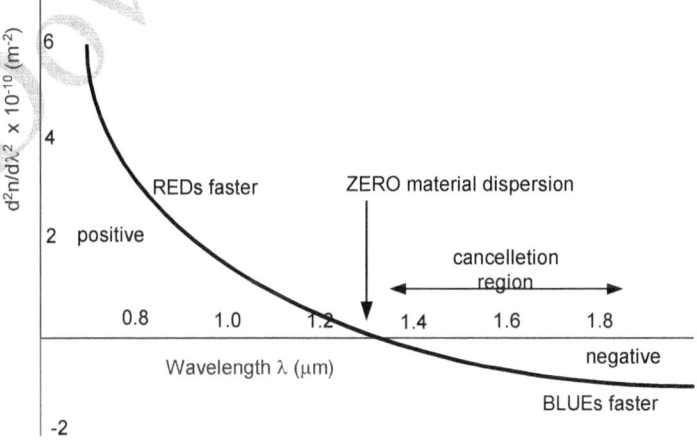

Figure 6   Material dispersion of SiO_2 glass. Zero occurs at 1.3 µm. RED is used for longer and BLUE for shorter wavelengths but all is in IR region.

Having so many effects turns out to be an advantage because single mode fibre eliminates the modal effect and the remaining chromatic (negative) and waveguide dispersion (positive) can be offset against each other at **1550 nm** (lowest loss in Figure 3A) to create **dispersion-shifted fibre** (see advanced fibres and Figure 13 later).

TABLE 1

Source	SI	GI	Monomode
LED λ=850 nm $\Delta\lambda$=50 nm	10/500 3	1000/500 3	100,000/500 2.5
Laser λ=850 nm $\Delta\lambda$=2 nm	10/25,000 3	1000/25,000 3	100,000/25,000 2.5
Laser l=1.3 μm $\Delta\lambda$=2 nm	10/100,000 1.0	1000/100,000 1.0	100,000/100,000 0.7
Laser l=1.55 mm DFB type	Library Exercise	Exercise	Exercise

Table 1 Dispersion and attenuation for various systems. The key to the data is: Modal dispersion/Material dispersion (both in MHz km) followed by Attenuation (dB km-1). The 1.55 μm laser TX case is left as an ambitious research exercise and can be done for monomode and dispersion shifted fibre.

Mode Cut-off Mechanisms

The guided LP modes may be investigated further with the help of Figure 2 that we used to find the turning points. We visualise the two graphs moving relative to each other and to the axis as shown below. The $n^2(r)k^2-\beta^2$ graph will move up as β reduces and down as β increases. This produces two extremes where the mode is said to be cut-off. An allowed range of β values therefore exists depending on profile n(r) and wavelength (contained in k).

The first case depicted in Figure 7a is where we consider modes with increasing ν. At a limiting value the ν^2/r^2 plot rises above the other and no turning point can be found. The annular ring has shrunk entirely. Such modes are not guided.

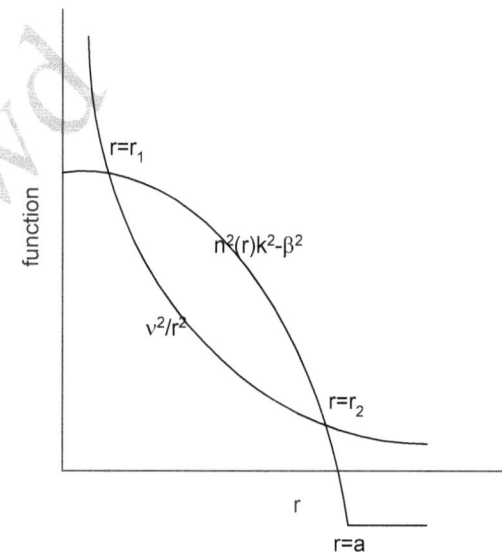

Figure 2 (Repeat) Finding the turning points r_1 and r_2 by the WKB method.

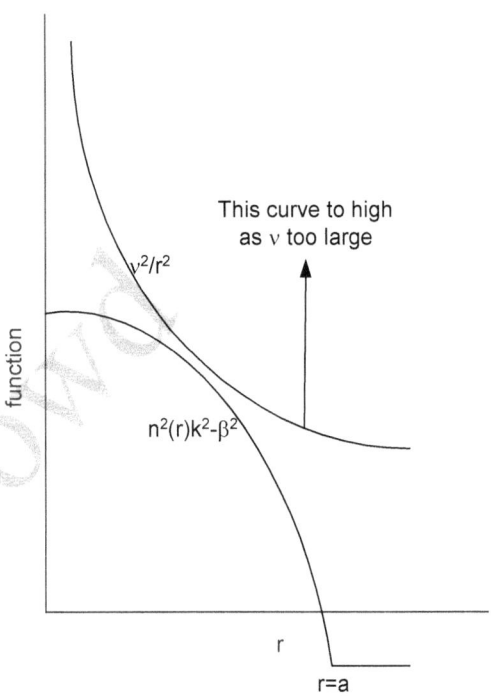

Figure 7a   Cut-off mechanism 1. The v^2/r^2 curve has moved up with increasing mode number v until no intersection occurs. Modes with that value v or greater cease to be guided.

In the next two cases the β value is allowed to rise to a maximum β_{max} and then fall to a minimum β_{min}. In Figure 7b the whole plot resembling profile squared has gone below axis so no turning points are possible. In Figure 7c that curve has risen entirely above axis and again such modes are cut-off. In this case another phenomenon appears; the third intersection r_3 means a new region to the right permits guided propagation and tunneling of light from the inner guidance region takes place. Such "evanescent modes" radiate energy to the cladding since r_3 exceeds a and after a few tens of metres and are said to be lossey.

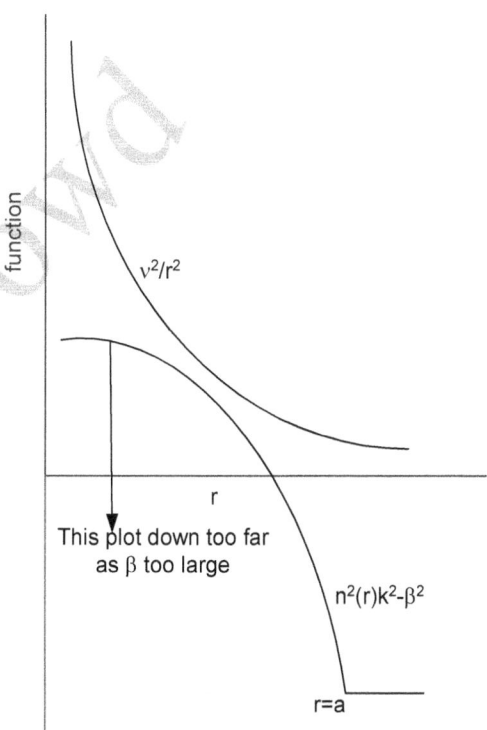

Figure 7b   Cut-off mechanism 2. The plot resembling the index profile squared moves down with increasing β until no intersection occurs. This situation must occur when the whole curve is below the axis and describes maximum β.

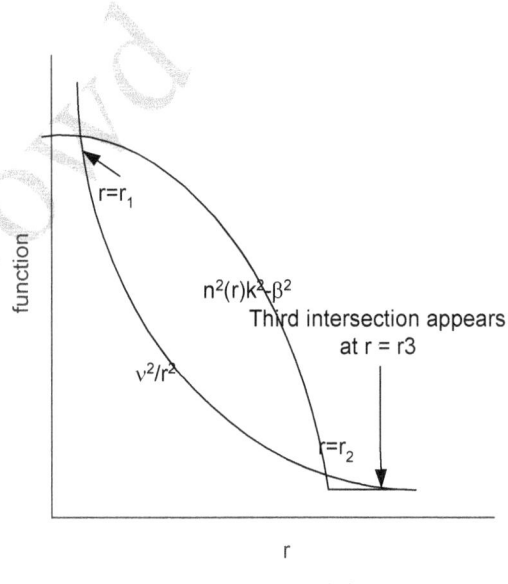

$$r=r_1$$

$$n^2(r)k^2-\beta^2$$

Third intersection appears
at r = r3

$$v^2/r^2$$

$$r=r_2$$

r

r=a

function

Figure 7c Cut-off mechanism 3. A third intersection r_3 occurs when the whole curve rises above the axis as when b decreases too much. This case describes minimum β.

You Do...

Ex 16: By studying the figures above for cut-off and imagining the graphs moving up or down so that all the curve is above or all is below the axis...

Show $\qquad$ $\beta_{min} = n_2 k$ $\qquad$ **cut-off condition**

Show $\qquad$ $\beta_{max} = n_1 k$ $\qquad$ **cut-off condition**

Hence $\qquad$ $n_2 k \;<\; \beta \;<\; n_1 k$ $\qquad\qquad$ (21)

Ans Ex 16 at end handbook.

Number of Guided Modes

The LP modes of the fibre carry power and since they are eigen-solutions to the wave equation they all have equal energy. To see how much total energy we must address the question of how many modes exist in a multimode fibre?

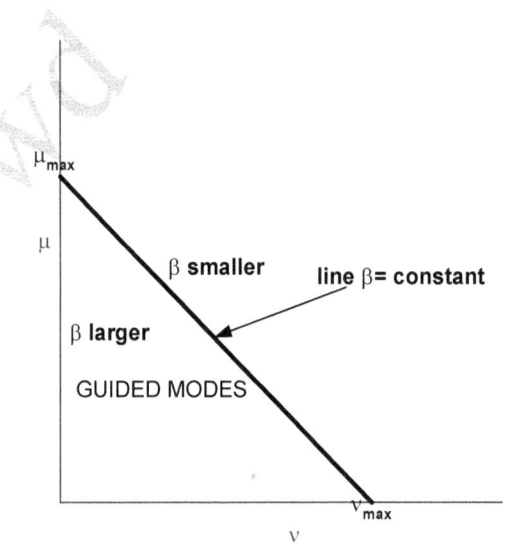

Figure 8 Guided LP modes depicted in mode number space.

Since each LP mode is defined by two integers we can draw a 2D chart, Figure 8, which depicts each mode as a dot with its ν number on the horizontal axis and μ number on the vertical. There turns out to be a line above which the mode does not exist as the integers rise above an allowed limit for propagation constant β. Recall we have already found the permissible range of β.

When we now count the dots under that line the result is the number of modes in the guide.

That count is essentially the same as the area under the curve which we get by integration. Hence the number M of modes, with β in the permitted range, equals the integral of vertical μ with respect to horizontal v:

$$M(\beta) \approx 4\int_0^{v_{max}} \mu dv$$

$$= \frac{4}{\pi}\int_0^{v_{max}} \int_{r_1}^{r_2} \frac{1}{r}\left\{\left[n^2(r)k^2 - \beta^2\right]r^2 - v^2\right\}^{\frac{1}{2}} drdv \qquad (22)$$

But for completeness we have included a *multiplier 4* since there are both sine and cosine possibilities that are equally valid or periodic at the starting point Eq (2) and furthermore there can be two polarisation states of the same mode. So the integral in Eq (22) is immediately expanded by substituting for μ using the WKB eigenvalue Eq (16).

Using $n(R)k = \beta$ and keeping in mind the range for β given by Eq (21), namely $n_2k < \beta < n_1k$, we get:

$$M(\beta) = \int_0^R \left[n^2(r)k^2 - \beta^2\right]rdr \qquad (23)$$

Recall $\qquad n(r) = \left[1 - 2\Delta\left(\frac{r}{a}\right)^{\gamma}\right]^{\frac{1}{2}} \quad \cdots\cdots \quad r < a$

$\qquad\qquad\qquad = n_1[1 - 2\Delta]^{\frac{1}{2}} \quad \cdots\cdots \quad r \geq a$ $\qquad (24)$

with $\qquad \Delta = \left(n_1^2 - n_2^2\right)/2n_1^2$ $\qquad\qquad\qquad (25)$

Here we must square n(r), so by Eq (24) the exponent becomes 2γ, and then integrated (see Ex 16 below and end of book) so that

$$\Rightarrow \qquad M(\beta) = \frac{\gamma}{\gamma+2}(n_1 ka)^2 \Delta \left[\frac{n_1^2 k^2 - \beta^2}{2n_1^2 k^2 \Delta}\right]^{\left(\frac{\gamma+2}{\gamma}\right)} \qquad (26)$$

$$\Rightarrow \qquad \beta = n_1 k \left[1 - 2\Delta \left(\frac{M(\beta)}{N} \right) \right] \qquad (27)$$

$$N = \left[\frac{\gamma}{(\gamma + 2)} \right] (n_1 ka)^2 \Delta \qquad (28)$$

Here we have renamed the general M(β) simply as N, the number of modes below the line for the limiting value of β and therefore the total number.

Let us for convenience define the fibre **V-parameter**:
$$\left(V \equiv n_1 ka \sqrt{2\Delta} \right)$$

This along with eq. (28) $\Rightarrow \qquad N = \frac{\gamma}{2(\gamma + 2)} V^2 \qquad (29)$

The last equation now provides tidy expressions for number of modes for various profiles:

$$\left. \begin{array}{ll} \gamma \rightarrow \infty \quad (step-index) \quad \cdots \quad N = \frac{1}{2} V^2 \\[2mm] \gamma = 2 \quad (parabolic) \quad \cdots \quad N = \frac{1}{4} V^2 \end{array} \right\} \qquad (30)$$

Observe that our optimised GI fibre with parabolic profile has only half as many propagating modes. Since the modes are eigen-solutions to the wave equation they carry equal energy. Hence only half as much power is carried by the GI fibre. That is the price we must pay for 100,000 times higher bandwidth.

You Do...

Ex 17: Square Eq (24), insert result into Eq (23) and use the known max and min values for β to produce, after integration, Eq (25).

Hint: reverse the order of integration and use the tabulated intgrand of arc-sine from standard integral tables.

Answer Ex 17 at end handbook.

LP or Linearly Polarised Modes in SI Fibre

A deeper examination reveals that each LP mode is in fact the superposition of two waves, one with strong magnetic (HE) and the other strong electric character (EH) and that the mode numbers for these are related to those of our selected LP mode as shown below:

$$HE_{v+1,\mu} \quad + \quad EH_{v-1,\mu} \quad \Rightarrow \quad LP_{v\mu}$$

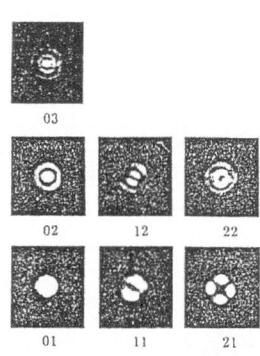

03

02   12   22

01   11   21

Figure 4 (Repeat). The LP modes as measured by Stolen.

Recall Eq.(2) again, for LP modes:

$$\psi = F(r)\cos(v\phi)e^{i(\omega t - \beta z)} \tag{31}$$

As before

$$\frac{d^2 F}{dr^2} + \frac{1}{r}\frac{dF}{dr} + \left(K^2 - \frac{v^2}{r^2}\right)F = 0 \tag{32}$$

Where now we have defined

$$K^2 = n^2(r)k^2 - \beta^2 \tag{33}$$

The form of Eq.(32) is well known in mathematics and its solutions are cylinder functions. This does not surprise us since our light guide is cylindrical. The known cylinder functions are called Bessel and

Hankel. Of these two the Bessel function $J_v(Kr)$ remains finite at the co-ordinate origin $r = 0$ and is therefore the appropriate cylinder function solution to eq.(32) for the <u>core</u> region:

$$F(r) = AJ_v(\kappa.r) \quad \cdots \quad |r| < a \tag{34}$$

For the special case of <u>step index or SI fibre</u> we will use for the core region $\kappa = K$ *with* $n(r) = n_1$ in eq.(33) so that we now get

$$\kappa = \left(n_1^2 k^2 - \beta^2\right)^{1/2} \tag{35}$$

For the <u>cladding</u> (postulate infinite extent) the solution is different. Here $n_2 < n_1$ and $n_2 k < \beta < n_1 k$ for guided modes. Introduce $\gamma = K/i$ (not shape factor!!) so

$$\gamma = \left(\beta^2 - n_2^2 k^2\right)^{1/2} \tag{36}$$

and the cylinder function must have imaginary argument since the bracket term must be negative.

Furthermore for guided modes $F(r)$ vanishes as $r \to \infty$. These two characteristic properties indicate the modified Hankel function $K_v(\gamma r)$ to describe the cladding field:

$$F(r) = BK_v(\gamma r) \quad \cdots \quad |r| > a \tag{37}$$

Note $\qquad \kappa^2 + \gamma^2 = \left(n_1^2 - n_2^2\right)k^2 \tag{38}$

Compare $\qquad V = ka\left(n_1^2 - n_2^2\right)^{1/2} \tag{39}$

This latter is the so-called V-parameter for the optical fibre.

$$\Rightarrow \quad \left[\left(\kappa a\right)^2 + \left(\gamma a\right)^2 = V^2\right] \tag{40}$$

65

At the core-cladding discontinuity the field and its derivative are continuous:

$$
\left.
\begin{array}{rcl}
AJ_v(\kappa.a) & - & BK_v(\gamma a) = 0 \\
\kappa AJ_v'(\kappa.a) & - & \gamma BK_v'(\gamma a) = 0
\end{array}
\right\} \tag{41}
$$

Solution by determinant method:

$$
\kappa J_v'(\kappa.a)K_v(\gamma a) = \gamma J_v(\kappa.a)K_v'(\gamma a) \tag{42}
$$

We may use from the mathematical handbook known relations for cylinder functions Eq (43):

$$
\left.
\begin{array}{rcl}
J_v'(x) & = & \dfrac{v}{x}J_v(x) - J_{v+1}(x) \\[2mm]
and \quad K_v'(x) & = & \dfrac{v}{x}K_v(x) - K_{v+1}(x)
\end{array}
\right\} \tag{43}
$$

The previous equation (42) now becomes

$$
\left[\; \kappa J_{v+1}(\kappa.a)K_v(\gamma a) = \gamma J_v(\kappa.a)K_{r+1}(\gamma a) \;\right] \tag{44}
$$

This equation (44) is the eigenvalue equation for the LP modes.
It is used simultaneously with equation (40), where (γa) and $(\kappa.a)$ are variables while V is the fibre parameter and v is the mode number, to produce the propagation parameter β for that mode. The last step also requires equations (35) and (36).

You Do...

Ex 18: Show that Eq(40) and Eq(44) solved simultaneously with the help of Eq(35) and Eq(36) can deliver the propagation constsant β for a selected LP mode.

Ans Ex 18 at end handbook.

o

Single-mode Fibres

Cylinder function computer sub-routines permit solution to the LP mode eigenvalue equation (44) by numerical methods. The result is the propagation constant β for each permitted mode and the spread of these represents the modal dispersion in a SI fibre. As the guide's core-diameter is reduced the number of permitted modes falls according to Eq(30) $N = V^2/2$ and we would expect that for $V^2 = 2$ or $V = 1.414$ the result should be a lone mode or $N = 1$.

In fact the step-index or S.I. fibre becomes single mode (derived below) when

$$V < 2.405 \qquad\qquad (45)$$

The reason for this paradox is that the starting assumption to derive Eq(30) had a slight simplification that does not matter when there are hundreds of modes as happens with typical multimode operation. We simplified $2\mu+1$ to 2μ. In the single mode case that assumption is invalid.

Mode Cut-off and the V-parameter

The cut-off condition whereby the argument in equation (36) turns from imaginary to real and the energy gets radiated instead of guided is

$$\gamma = 0 \tag{46}$$

Hence equation (40)

$$V^2 = (\kappa . a)^2 + (\gamma a)^2$$

gives

$$V_c = \kappa . a \tag{47}$$

Taking this as argument for the Bessel function in the eigenvalue eq.(44) and assuming $\gamma a \ll 1$ in the modified Hankel function approximations I and II (also from the mathematical handbook)...

$$
\left.
\begin{array}{ll}
I \quad & K_\nu(x) = \ln\!\left(\dfrac{2}{x}\right) \quad \cdots \; \nu = 0 \\[2ex]
and \; II \quad & K_\nu(x) = \dfrac{(\nu-1)!}{2}\left(\dfrac{2}{x}\right)^2 \quad \cdots \; \nu \geq 1
\end{array}
\right\} \tag{48}
$$

We obtain using I:

$$I \qquad J_1(V_c) = J_0(V_c)/\left[V_c \ln\!\left(\dfrac{2}{\gamma a}\right)\right]$$

(49) Hence $\qquad J_1(V_c) = 0$ as $\ln(\text{infinity}) = 0$. $\tag{50}$

This is the cut-off condition for modes with $\nu = 0$

When $\nu \geq 1$ we obtain using II:

$$II \qquad \dfrac{2\nu}{V_c} J_\nu(V_c) - J_{\nu+1}(V_c) = 0 \tag{51}$$

Use from the mathematical handbook the functional relation:

$$V_c J_{v-1}(V_c) + V_c J_{v+1}(V_c) = 2v J_v(V_c) \tag{52}$$

$$\Rightarrow \quad J_{v-1}(V_c) = 0 \quad \text{cut-off condition for modes with } v \geq 1 \tag{53}$$

For the LP_{11} mode in particular this indicates $\quad J_0(V_c) = 0$

$$\underline{V_c = 2.405} \tag{54}$$

This is telling us that the LP_{11} mode becomes cut-off as the V-parameter is reduced below this critical value 2.405 and therefore only the LP_{01} mode remains giving us single mode operation or monomode fibre.

We defined the fibre V-parameter $\quad \left(V \equiv n_1 ka\sqrt{2\Delta}\right) \quad$ so that by reducing the fibre diameter 2a sufficiently we must arrive at single mode propagation. Inserting V = 2.405 and typical n_1, n_2, and λ values in that definition will show that the fibre is monomode below about 10 μm diameter.

You Do...

Ex 19: From V parameter definition and knowing the associated single mode condition estimate the radius of the fibre where
$\lambda = 1.5$ μm, $n_1 = 1.5$, $n_2 = 1.49$

Ans Ex 19 at end handbook.

○

The single mode fibre radius answer to Ex 19 is 3.3 μm or 6.6 μm diameter. Compare multimode fibre with 50 μm core diameter. By varying Δ the monomode answer can be up to 10 μm.

Cut-off V-parameter

The above discussion was for SI fibre but the whole analysis from Eq(32) to here would need to be repeated for other index profiles each with their own shape factor γ. In fact computational methods would be required for this due to the complexity when the core index is no longer fixed at a constant n_1. Nonetheless this has been performed and the result is plotted in Figure 9.

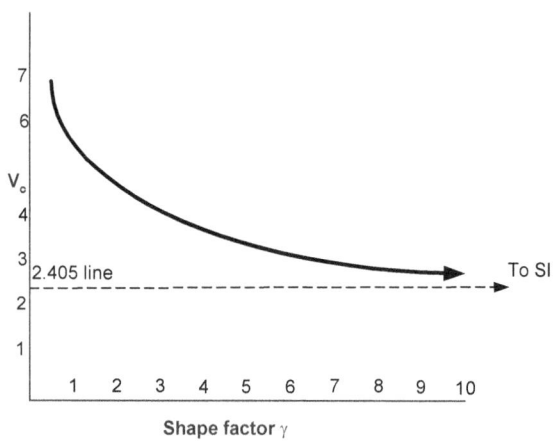

Figure 9. Cut-off V-parameter versus profile shape-factor; SI fibre to right.

Cut-off Wavelength of the Fibre

The wavelength λ is itself contained within the definition of V-parameter so that we should expect that at a certain wavelength the V-parameter ensures that the mode becomes cut-off. We find that after designing the fibre to be monomode and then by using shorter and shorter wavelength lasers ultimately a second mode appears. That is called the *cut-off wavelength* because at longer than that λ the second mode gets eliminated. This λ_c can be measured by launching variable wavelength light from a monochromator into a reel of fibre and observing the output light pattern with a IR camera. At longer than cut-off wavelength there is only the LP_{01} mode but at shorter than λ_c the LP_{11} mode appears and the centre of the optical near-field dims as depicted in Figure 10.

Figure 10(below). As the injected light wavelength shortens below cut-off wavelength a second mode appears so the pattern is a superposition of LP_{01} and LP_{11}.

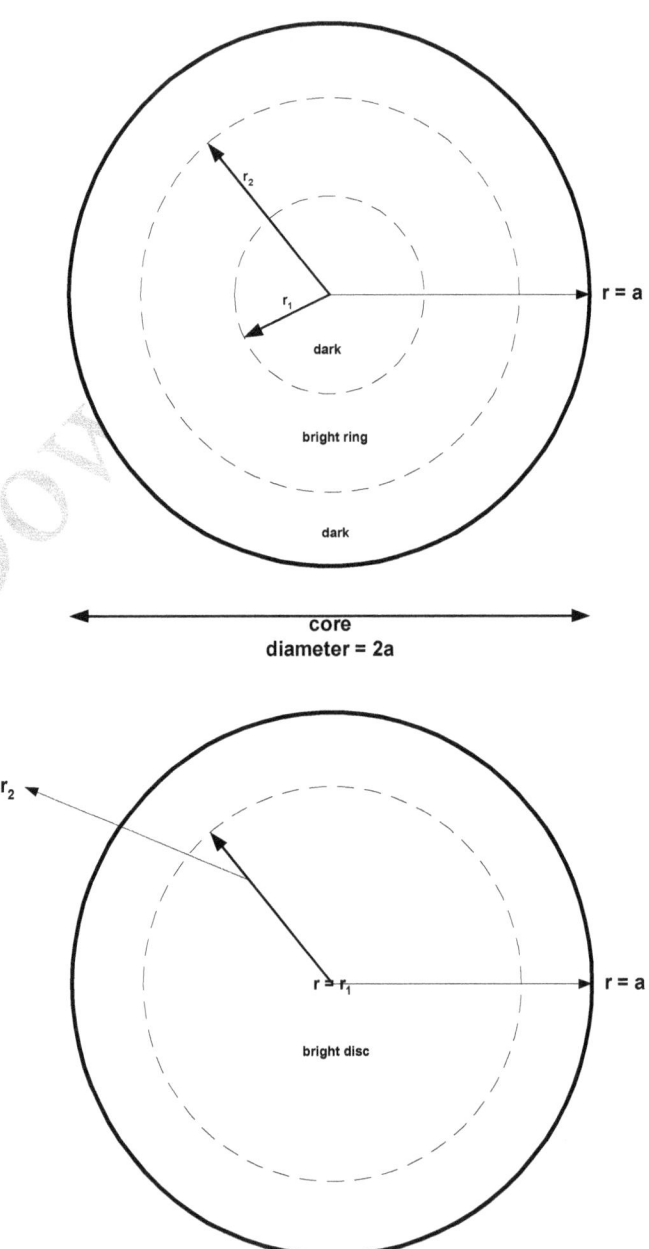

r_2

r_1

dark

bright ring

dark

r = a

core
diameter = 2a

r = r$_2$

r = r$_1$

r = a

bright disc

72

5 Advanced Fibre Designs

We have learned to design the ultimate single mode fibre at a selected wavelength but when a range of many wavelengths are used simultaneously as separate channels in the one fibre the system is called WDM for wavelength division multiplexing. In such cases only one of the comb of channels would be optimised while the others would suffer. Therefore we need to return to the analysis of trade-offs to see how the full set of wavelengths might be made to propagate with an equal but minimal dispersion rather than just one being optimised while all others degrade. The outcome is a more advanced index profile such as depicted for comparison in Figures 11 and 12. These are the result of very detailed mathematical scrutiny and are deployed in dense WDM systems. The whole C-band in the 1550 nm region of the optical spectrum is filled with channels 50 GHz apart in optical frequency. This gives the so called ITU (International Telecommunications Union) comb of channels.

Figure 11. Non-dispersion-shifted profile and triangular profile *zero* dispersion-shifted fibre optimised at low-loss wavelength 1550 nm rather than prior minimum-dispersion λ at 1300 nm

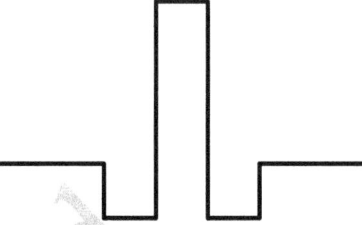

Figure 12   W-profile for dispersion-flattened fibre accommodating a comb of C-band channels with a compromise dispersion value that is flat across all chosen wavelengths.

Polarisation Maintaining Fibre

There a many situations where it is advantageous to deploy fibre that transmits one polarisation only.   For example sensors often utilise polarised light. Coherent communications (like optical FM) systems are another case in point. By creating more stress across the diameter in one preferred direction it is found that a single polarisation is suppressed with the other remaining. That stress is achieved in a number of ways depicted in Figures 13 and 14.

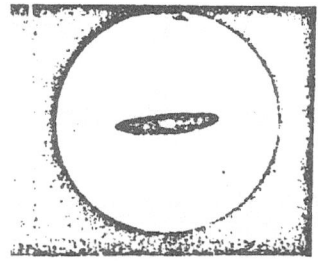

Figure 13.   Polarisation maintaining fibres; elliptical core.

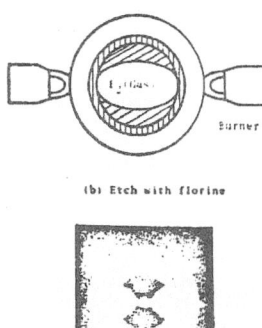

(b) Etch with florine

(d) Final cross-section

Figure 14. Polarisation maintaining fibre with bow-tie core. The perform growth process is shown revealing the etching stage with burner and fluorine gas to create the stress-creating shaped core. (University of Southampton).

In one case the glass preform is grown with an elliptical core and in the other with a bow-tie shape. This causes the asymmetrical stress required.

We can represent the optical power flowing down the z-axis as comprising x and y polarisations orthogonal to one another and to the axis. The rate of transfer of power from P_x to P_y take place with distance z at a rate we will designate h so that the equations (55) summarise what is happening.

$$\frac{dP_x}{dz} = h\left(-P_x + P_y\right) \quad \text{and} \quad \frac{dP_y}{dz} = h\left(P_x - P_y\right) \tag{55}$$

$$\cdots \quad \eta = \tanh(hz) \tag{56}$$

The form of these coupled equations is familiar in mathematics where the solution takes the form of equation (56) and the hyperbolic tangent trend becomes evident.

Alternative Glasses

The conventional optical glass is SiO2, namely silica or silicate but other glasses have been explored to see if longer links could be created without expensive repeaters. One such candidate was fluoride glass and its performance is shown in Figure 17 for comparison. While promising in the laboratory the fluoride glass is however unstable in air and not used for that reasons in broadband cables.

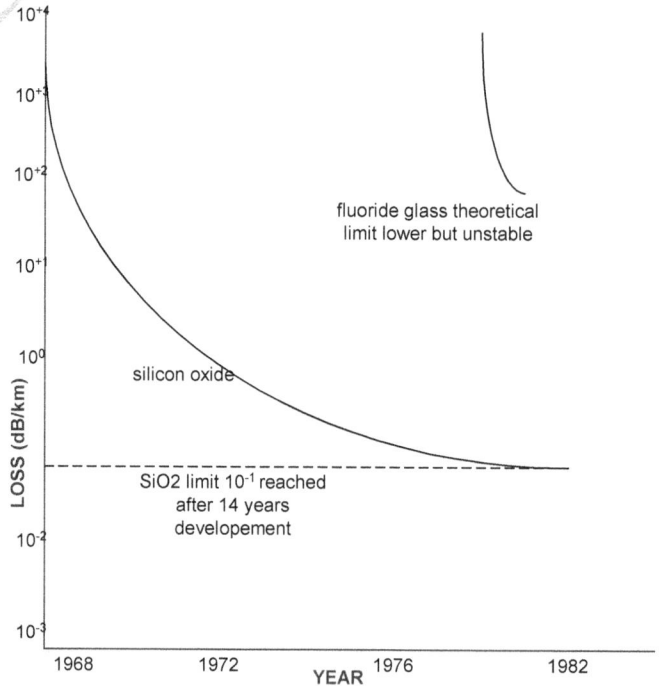

Figure 17. Attenuation (loss) for oxide and fluoride glass.

Summary 4

Dispersion in fibres is analysed generally but especially for the modal effects and the range of propagation parameters was found:

$$n_2 k \quad < \quad \beta \quad < \quad n_1 k$$

These limits are useful to compute number of guided modes for SI and GI profiles. The SI case was further studied to produce the eigenvalue equation for the β associated with each particular $LP_{\nu\mu}$ mode via Eq (44) along with Eq(40). Then the special case of SI monomode propagation was considered and the V-parameter below which that happens was derived using known cylinder function relations. Advanced profiles for WDM systems were discussed and polarisation maintaining designs were reviewed.

Answers to Ex 10-19

Ans Ex 10

Using given values: $n_1 - n_2 = 1.51 - 1.50 = 0.01$

$r = 25 \ \mu m$ $\lambda = 1.55$ mm

$r/\lambda = 16$ approx so index variation over 1λ is $0.01/16 = 0.0006$

This is negligeable in the context of $n_1 = 1.5$ the assumption in valid.

Ans Ex 11

$$\psi = F(r)\cos(v \, \phi)e^{i(\omega t - \beta z)}$$

$$\frac{d\psi}{dr} = \frac{dF(r)}{dr}\cos v\phi \, \exp i(\omega t - \beta z)$$

$$\frac{\delta^2\psi}{\delta r^2} = \frac{d^2 F(r)}{dr^2}\cos v\phi \, \exp i(\omega t - \beta z)$$

$$\frac{\delta^2\psi}{\delta \phi^2} = F(r)v^2 \cos v\phi \, \exp i(\omega t - \beta z)$$

$$\frac{\delta^2\psi}{\delta z^2} = (i\beta)^2 F(r)\cos v\phi \, \exp i(\omega t - \beta z)$$

Adding all according to cylindrical wave Eq(2) and cancelling   term $\cos v\phi \, \exp i(\omega t - \beta z)$ **across gives us using on right hand side k=$2\pi/\lambda$ with** $\omega = 2\pi/v$ **and from end of last section 2 that c^2=1/$\mu_0\varepsilon_0$:**

$$\frac{\delta^2\psi}{\delta r^2} + \frac{1}{r}\frac{\delta\psi}{\delta r} + \frac{1}{r^2}\frac{\delta^2\psi}{\delta\phi^2} + \frac{\delta^2\psi}{\delta z^2} = n^2\varepsilon_0\mu_0\frac{\delta^2\psi}{\delta t^2} \tag{2}$$

$$\Rightarrow \quad \frac{d^2 F}{dr^2} + \frac{1}{r}\frac{dF}{dr} + \left(n^2 k^2 - \beta^2 - \frac{v^2}{r^2}\right)F = 0 \tag{3}$$

Ans Ex 12

$$\frac{d^2F}{dr^2} + \frac{1}{r}\frac{dF}{dr} + \left(n^2k^2 - \beta^2 - \frac{v^2}{r^2}\right)F = 0 \tag{3}$$

Transform from linear r-space to log x-space using r = ae^x

Hence dr = ae^x dx = rdx and dx = (1/r)dr

Eq(3) times r² is starting step:

$$\Rightarrow \quad r^2\frac{d^2F}{dr^2} + r\frac{dF}{dr} + \left(k^2r^2 - v^2\right)F = 0 \quad \text{where} \quad k^2 = n^2k^2 - \beta^2 \tag{a}$$

$$dF = \frac{dF}{dx}dx = \frac{dF}{dx}\frac{1}{r}dr$$

$$\therefore \frac{dF}{dr} = \frac{1}{r}\frac{dF}{dx}$$

$$\frac{d^2F}{dr^2} = \frac{-1}{r^2}\frac{dF}{dx} + \frac{1}{r}\frac{d}{dr}(\frac{dF}{dx}) = \frac{-1}{r^2}\frac{dF}{dx} + \frac{1}{r^2}\frac{d^2F}{dx^2}$$

$$r^2\frac{d^2F}{dr^2} = -\frac{dF}{dx} + \frac{d^2F}{dx^2}$$

$$\therefore r^2(\frac{d^2F}{dr^2}) + r\frac{dF}{dr} = \frac{d^2F}{dx^2}$$

Hence Eq(a) above becomes:

$$\Rightarrow \quad \frac{d^2F}{dx^2} + \left(k^2r^2 - v^2\right)F = 0$$

$$\Rightarrow \quad \frac{d^2F}{dx^2} + \left(k^2a^2e^{2x} - v^2\right)F = 0$$

Ans Ex 13

Since $F(r) = A(r)e^{is(r)}$ describes the radial amplitude and phase properties:

$$\psi = F(r)\cos(\nu\,\phi)e^{i(\omega t - \beta z)} = A(r)e^{is(r)}\cos(\nu\,\phi)e^{i(\omega t - \beta z)} \qquad \text{MODE}$$

Now use exponential to trigonometric conversion $\cos\nu\phi = \tfrac{1}{2}(e^{i\nu\phi}\,e^{-i\nu\phi})$:

$$\therefore \psi = \frac{1}{2}A(r)\{\exp[i(\omega t - \beta z - \nu\phi + S(r))] + \exp[i(\omega t - \beta z + \nu\phi + S(r))]\} \qquad (13)$$

Ans Ex 14

Recall Eq(11)

$$S(r) = \int_1 \left[(n^2 k^2 - \beta^2) - \frac{\nu^2}{r^2}\right]^{\frac{1}{2}} dr \qquad (11)$$

However we only used the positive square root term on first visit therefore for a complete solution the negative S(r) version needs to be added in:

$$\therefore \psi = \frac{1}{2}A(r)\{\exp[i(\omega t - \beta z - \nu\phi + S(r))] + \exp[i(\omega t - \beta z + \nu\phi + S(r))]\}$$

$$+ \frac{1}{2}A(r)\{\exp[i(\omega t - \beta z - \nu\phi - S(r))] + \exp[i(\omega t - \beta z + \nu\phi - S(r))]\}$$

Finally revert to trig from exp with a reverse conversion to tidy up:

$\cos\nu\phi = \tfrac{1}{2}(e^{i\nu\phi} + e^{-i\nu\phi})$ and $\cos S(r) = \tfrac{1}{2}(e^{iS(r)} + e^{-iS(r)})$:

$$\therefore \psi = A(r)e^{is(r)}\cos(\nu\,\phi)\cos[S(r)]e^{i(\omega t - \beta z)} \qquad (14)$$

Ans Ex 15

Eq(19) $\left[\Delta\tau_{opt} \approx \dfrac{Ln_1}{8c}\Delta^2 \right]$ for optimised-profile modal dispersion with typical

fibre values gives using 1 km span:

$$\left[\Delta\tau_{opt} \approx \frac{Ln_1}{8c}\Delta^2 = 1.5(\frac{1.5-1.49}{1.5})^2\frac{1}{(24)10^8} \right]$$

This is 0.03 ps/km for GI parabolic index profile which is hundreds of thousands times improvement on SI fibre. Hence the deep cusp at $\gamma = 2$.

Ans Ex 16

Cut-off max and min values for β are found by letting β get so large it pulls the plot in Figure (7) entirely below the axis. The expression $n_1^2(r)k^2-\beta^2 = 0$ as the index peak crosses the axis where $n(r)$ is n_1.

Hence $\beta_{max} = n_1 k$

In the reverse situation the plateaux for cladding where $n(r) = n_2$ passes the axis as b is so small giving $n_2^2(r)k^2-\beta^2 = 0$

Hence $\beta_{min} = n_2 k$

Ans Ex 17

Number of guided modes.

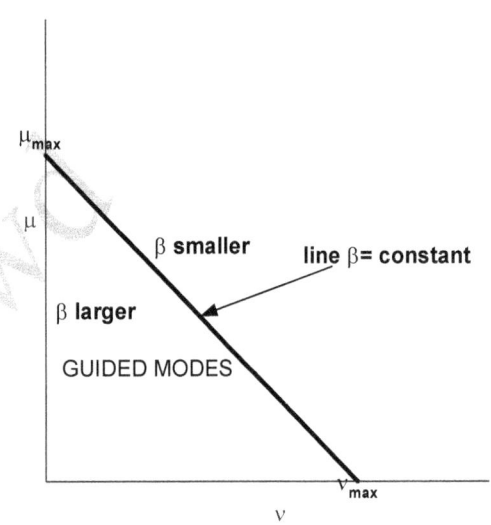

Figure 8   Guided LP modes depicted in mode number space

Since each LP mode is defined by two integers we can draw a 2D chart, Figure 8, which depicts each mode as a dot with its ν number on the horizontal axis and μ number on the vertical. There turns out to be a line above which the mode does not exist as the integers rise above an allowed limit for propagation constant β. Recall we have already found the permissible range of β.

When we now count the dots under that line the result is the number of modes in the guide.

Observe that the count is essentially the same as the area under the curve which we get by integration. Hence the number M of modes, with β in the permitted range, equals the integral of vertical μ with respect to horizontal $d\nu$:

$$M(\beta) \approx 4\int_0^{\nu_{max}} \mu d\nu$$

$$= \frac{4}{\pi}\int_0^{\nu_{max}} \int_{r_1}^{r_2} \frac{1}{r}\left\{\left[n^2(r)k^2 - \beta^2\right]r^2 - \nu^2\right\}^{\frac{1}{2}} drd\nu \qquad (22)$$

Here we substituted the expression for μ from the WKB analysis; that is the inner integral divided by π (strictly that gives $\mu+1/2$ but we assume μ is large).

But for completeness we have included a *multiplier 4* since there are both sine and cosine possibilities that are equally valid or periodic at the starting point Eq (2) and furthermore there can be two polarisation states of the same mode. So the single integral in Eq(22) is immediately expanded to double integral by substituting for μ using the WKB eigenvalue Eq (16)

$$S(r) = \int_{r_1}^{r_2}\left[n^2(r)k^2 - \beta^2 - \frac{\nu^2}{r^2}\right]^{\frac{1}{2}} dr = \left(\mu + \frac{1}{2}\right)\pi = \mu\pi \qquad (16)$$

Observe that in the Figure above ν is max when μ is zero and r_1 goes to zero because that is the LP mode with no radial repeats in the pattern.

Also as the rhs is now zero in Eq(16) the square bracket on lhs is zero with β minimised at $\beta_{min} = n_2 k$

$$\left[n^2(r)k^2 - \beta_{min}^2 - \frac{\nu_{max}^2}{r^2}\right] = 0$$

Hence:

$$\frac{\nu_{max}^2}{r^2} = n^2(r)k^2 - \beta_{min}^2 = n^2(r)k^2 - n_2^2 k^2 \qquad A$$

Profile:

$$n(r) = \left[1 - 2\Delta \left(\frac{r}{a} \right)^{\gamma} \right]^{\frac{1}{2}} \quad \cdots\cdots \quad r < a$$
$$n_2 = n_1 \left[1 - 2\Delta \right]^{\frac{1}{2}} \quad \cdots\cdots \quad r \geq a \qquad\qquad (17)$$

Hence

$$n^2(r) - n_2^{\,2} = \left[2n_1^{\,2}\Delta \left(1 - \frac{r}{a} \right)^{\gamma} \right] \qquad\qquad \mathbf{B}$$

Sub this Eq B into A above gives

$$v_{max}^{\,2} = 2r^2 n_1^{\,2} k^2 \Delta \left[\left(1 - \frac{r}{a} \right)^{\gamma} \right] \qquad\qquad \mathbf{C}$$

In Eq(22) as the range is now 0 to $v_{\mu\alpha\xi}$ and r goes 0 to a to capture all modes possible let us switch to N for mode number.

Using $n(R)k = \beta$ and keeping in mind the range for β given by Eq (21), namely $n_2 k < \beta < n_1 k$, we get using Eq A above:

$$N = \frac{4}{\pi} \int_0^{r_{max}} \int_0^{a} \frac{1}{r} \left\{ \left[n^2(r)k^2 - \beta_{min}^{\,2} \right] r^2 - v^2 \right\}^{\frac{1}{2}} dr\,dv \qquad (23)$$

Reverse order of integration:

$$N = \frac{4}{\pi} \int_0^{a} \frac{1}{r} \int_0^{v_{max}} \left[v_{max}^{\,2} - v^2 \right]^{\frac{1}{2}} dv\,dr \qquad (24)$$

The integral with respect to v is well known as an arcsine functional relation in the mathematical handbook (or using integration by parts) producing a $\pi/4$ that then cancels by good fortune the $4/\pi$ with the result:

$$N = \frac{4}{\pi} \int_0^{a} \frac{1}{r} \left[\frac{\pi v_{max}^{\,2}}{4} \right] dr = \int_0^{a} \frac{1}{r} \left[v_{max}^{\,2} \right] dr$$

But according to C: $\quad v_{max}^{\,2} = 2r^2 n_1^{\,2} k^2 \Delta \left[\left(1 - \frac{r}{a} \right)^{\gamma} \right]$

$$N = \int_0^{a} \frac{1}{r} 2r^2 n_1^{\,2} k^2 \Delta \left[\left(1 - \frac{r}{a} \right)^{\gamma} \right] dr$$

Thus we get an integrand a sole function of r. Here we must bring r inside the bracket so the , so by Eq (24) the exponent becomes $\gamma+1$, and then integrated so that the power becomes $\gamma+2$:

$$N = 2n_1^2 k^2 \Delta \left[\left(\frac{r^2}{2} \right) - \frac{1}{a^\gamma} (\frac{r^{g+2}}{\gamma^2}) \right]$$ and this definte integral must be

evaluated between limits r=0 to r=a

giving $$N = 2n_1^2 k^2 \Delta a^2 \left[\frac{1}{2} - \frac{1}{\gamma+2} \right]$$

Finally: $$N = \left[\frac{\gamma}{(\gamma+2)} \right] (n_1 ka)^2 \Delta \qquad (28)$$

Here we have renamed the general $M(\beta)$ simply as N, the number of modes below the line for the limiting value of β and therefore the total number.

Let us for convenience define the fibre V-parameter: $\left(V \equiv n_1 ka\sqrt{2\Delta} \right)$

This along with eq. (28) $\Rightarrow$ $$N = \frac{\gamma}{2(\gamma+2)} V^2 \qquad (29)$$

Ans Ex 18

$$\kappa = \left(n_1^2 k^2 - \beta^2\right)^{\frac{1}{2}} \tag{35}$$

$$\gamma = \left(\beta^2 - n_2^2 k^2\right)^{\frac{1}{2}} \tag{36}$$

$$\left[(\kappa a)^2 + (\gamma a)^2 = V^2\right] \tag{40}$$

$$\left[\ \kappa J_{v+1}(\kappa .a) K_v(\gamma a) = \gamma J_v(\kappa .a) K_{r+1}(\gamma a)\ \right] \tag{44}$$

Eq (40) and (44) have variables κa and γa but selected parameters V and a depending on the fibre chosen. Once κ and γ are found then they are inserted into Eq (35) and (36) to quantify β.

Ans Ex 19

$\left(V \equiv n_1 ka\sqrt{2\Delta}\right)$; $\Delta = (n_1 - n_2)/n_1 = 0.01/1.5$ and $k = 2\pi / \lambda$

Hence a = $2.405/2\pi(0.115)$ in microns

So a= 3.33 μm for the single mode fibre or diameter D= 6.6 μm.

This means single mode fibre is about 10 times thinner core than multimode but also the splicing and connectors must be ten times more precise. So we need top quality mechanical engineering and laser monitoring. We will need to use OTDRs (optical time domain reflectometers) to assess the precision of all joints.

Conclusion

We have discovered by mathematics how to create broadband over fibre.

Dispersion in its many guises was scientifically modelled to produce optimal refractive index profile design, then implemented by computer controlled chemical vapour deposition. May you enjoy the broadband communications that results from your new understanding of these "glassworks" of wonderful engineering.

Typical Examination Questions.

$$c = 3 \times 10^8 \quad ms^{-1}$$
$$e = 1.6 \times 10^{-19} \, C$$
$$h = 6.6 \times 10^{-34} \, Js$$
$$k = 1.38 \times 10^{-23} \, JK^{-1}$$
$$m_e = 9.11 \times 10^{-31} \, kg$$

Note: **Bessel and Hankel function relations are provided below at end paper.**

QUESTION A

(a) By considering *all* solutions to the following wave equation

$$\nabla^2 \psi = n^2 \varepsilon_0 \mu_0 \frac{\delta^2 \psi}{\delta t^2}$$

(1)

show that the *complete* expression for the guided mode is as equation (2) and that the corresponding eigenvalue equation for guided light in an optical fibre is given by equation (3):

$$\psi = A(r)\cos(\nu\phi)\cos\left[S(r)\right]e^{i(\omega t - \beta z)}$$

(2)

$$\int_{r_1}^{r_2} \left[n^2(r)k^2 - \beta^2 - \frac{\nu^2}{r^2} \right]^{\frac{1}{2}} dr = \left(\mu + \frac{1}{2} \right)\pi$$

(3)

50%

(b) Indicate briefly what further modification is taken into account before assessing the number of guided modes.

10%

(c) Having now solved for the guided modes state why equation (1) is so called.

This result contains the refractive index profile n(r). Outline how you would use computer methods to apply your result and discover numerically the optimum profile for multimode fibres and plot the expected outcome.

20%

(d) Outline how you would apply this result in an engineering context to manufacture fibres of the desired n(r) design.

20%

QUESTION B

Consider a general optical fibre index profile:

$$n(r) = n_1 \left[1 - 2\Delta \left(\frac{r}{a} \right)^s \right]^{\frac{1}{2}}$$

where s is the "shape factor" and other symbols have the usual meanings.

Given the eigenvalue equation for the LP modes:

$$\kappa J_{\nu+1}(\kappa a) K_\nu(\gamma a) = \gamma J_\nu(\kappa a) K_{\nu+1}(\gamma a)$$

where κ and γ depend on propagation constants and are related to the fibre V-parameter, show that the condition for single mode operation (i.e. cut-off condition) is:

$$V_C = 2.405$$

At what point in this argument is s value of infinity assumed?

Hence indicate briefly how the condition:

$$V_C = 4$$

might be arrived at as the corresponding condition for graded index fibre. Illustrate the trend for other profiles and discuss how and where this might find application in optical communication systems.

QUESTION C

(i)   Evaluate the maximum and minimum modal propagation constants β, in terms of free space wavelength and optical fibre parameters. Describe the mode cut-off mechanisms and explain the term "leaky mode".

(ii)   What is meant by "cut-off wavelength" and describe how it can be measured. Describe how this is related to core dimension or other fibre properties.

QUESTION D

Show that in a typical optical fibre refractive index profile design for telecommunications the basic starting assumption of the WKB theory will apply.

Establish the phase conditions for the guided modes (multimode case) and hence the **"turning point"** condition. Describe what happens in the various regimes for the modal propagation constant implied by this condition.

QUESTION E

(a)   Consider a general optical fibre index profile:

$$n(r) = n_1 \left[1 - 2\Delta \left(\frac{r}{a} \right)^s \right]^{\frac{1}{2}}$$

where s is the "shape factor" and other symbols have the usual meanings.

Given the eigenvalue equation for the LP modes:

$$\kappa J_{v+1}(\kappa a) K_v(\gamma a) = \gamma J_v(\kappa a) K_{v+1}(\gamma a)$$

Where κ and γ depend on propagation constants and are related to the fibre V-parameter, show that the condition for single mode operation (i.e. cut-off condition) is:

$$V_C = 2.405$$

(b)   At what point in the above argument is s = ∞ assumed?   Hence indicate briefly how the condition

$$V_C = 4$$

might be arrived at as the corresponding condition for graded index fibre.   Illustrate the trend for other profiles and discuss how and where this might find application in optical communication systems.

QUESTION F

Use the WKB theory to find the number of guided modes in an optical fibre with refractive index profile

$$n(r) = n_1 \left[1 - 2\Delta \left(\frac{r}{a} \right)^s \right]^{\frac{1}{2}}$$

where s is the "shape factor" and other symbols have the usual meanings.

Show that parabolic index fibre carries half the light of a step index fibre.

QUESTION G

(i)   Dispersion-shifted optical fibre has zero total dispersion at the loss minimum for the glass.
           Discuss how this arises and how it is achieved for the combined dispersion contributions.
           Describe further advanced refractive index profiles and their purpose.

(ii)   The loss mechanisms in the fibre produce the ultimate minimum in mid C-band.   Describe
           these and their respective spectra.

QUESTION H

A general index profile optical fibre is analysed by the WKB method and produces the expression

$$M(\beta_{min}) = \frac{4}{\pi} \int_{r=0}^{a} \frac{1}{r} \left[\frac{\pi^2 v_{max}^2}{4} \right] dr$$

for the number of guided modes, the symbols taking their normal meaning. Use this along with the Gloge and Marcatili expression for the index profile to show that approximately half as much energy may be carried by a graded-index fibre as a step-index fibre of corresponding dimensions.

QUESTION I

(i)   For the case of a multimode fibre the $LP_{\upsilon\mu}$ modes F(r) satisfy:

$$\frac{d^2 F}{dr^2} + \frac{1}{r} \frac{dF}{dr} + \left[n^2 \left(r \right) k^2 - \beta^2 - \frac{\upsilon^2}{r^2} \right] F = 0$$

The symbols take their usual meanings. By defining appropriate parameters κ and γ for the core and cladding regions respectively and then relating these to the V-parameter for the fibre, show that an eigenvalue equation for the LP modes can yield the modal propagation constants.

(ii)   Use the same eigenvalue equation to prove that in a step-index single-mode fibre the V-parameter associated with cut-off must be zero in order that the LP_{01} mode be leaky.

State the tendency towards this condition in terms of:

(a)   fibre radius,
(b)   transmission wavelength,
(c)   index profile.

Note: Bessel and Hankel function relations are provided below.

Mathematical properties used.

Exponential to trigonometric conversion:

$e^{\imath v \phi}$ = cosvϕ + isinvϕ

$e^{-\imath v \phi}$ = cosvϕ - isinvϕ

Cosvϕ = $\frac{1}{2}(e^{\imath v \phi} + e^{-\imath v \phi})$

Cylinder functions:

The Bessel function $J_v(Kr)$ remains finite at the co-ordinate origin $r = 0$ and is therefore the appropriate cylinder function solution to eq (32) for the fibre <u>core</u> region:

$$F(r) = AJ_v(\kappa r) \quad \cdots \quad |r| < a \qquad (34)$$

The modified Hankel function $K_v(\gamma r)$ is the cylinder function with imaginary argument and vanishes as $r \to \infty$. and applies to the <u>cladding</u>.

Functional relations for Hankel functions:

$$\left. \begin{array}{l} \text{I} \quad K_v(x) = \ln\left(\frac{2}{x}\right) \quad \cdots \quad v = 0 \\[2ex] \text{and II} \quad K_v(x) = \frac{(v-1)!}{2}\left(\frac{2}{x}\right)^2 \quad \cdots \quad v \geq 1 \end{array} \right\} \qquad (48)$$

Functional relation for three orders of Bessel function:

$$V_c J_{v-1}(V_c) + V_c J_{v+1}(V_c) = 2vJ_v(V_c) \qquad (52)$$

For the zero order Bessel function to be zero $J_0(V_c) = 0$ the argument is determined as the plot crosses the axis at 2.405:

$$\underline{V_c = 2.405} \qquad (54)$$

Hardware: Optosci Kit

See Optosci Manual and proceed with the selected hardware or HW experiments below.

HW-1
System losses and bandwidth.

Experiments 8.1, 8.2, 8.3 pp12-15 ED-COM Manual and B2 Measure fibre length by pulse transmission (Optosci Appendix B in the ED-COM Manual).

HW-2
Eye Diagrams and bit-error rate BER

See Fig 2.3, Eq (13), Fig 2.7,

Experiment 6.1   BER(COM) Optosci Manual pp 21
            6.2   22
            6.3   23
            6.4   24
            6.5   25

Name: _____ Student Number: _____

1.1 Draw the characteristic of output power against drive current by hand for LED and LASER. Then, mark the important points (bias point, threshold current).

LED

I [mA]									
P [uW]									

LASER

I [mA]									
P [uW]									

Graph:

2.1 Measure the output value for the patchcord (end-to-end), reel #1 and reel #2 for LED and LASER. Then, calculate the attenuation of link and compare. Make a comparison and evaluation.

Link/Source	P [uW]

Other parameters:

2.2 Calculate the length of fibre for each reel from the time difference value of biased signal and detecting signal after the fibre link. Do it for LED and LASER source. Consider results.

3.1 Modulate the LED source by square signal. Then, find the pulse risetime (10% to 90%) on the oscilloscope. Take measurements of both reel, interconnected reels and end-to-end scheme. Consider system and fibre properties. Compare the values Bandwidth-length product and bitrate-distance product.

Setup	U_{10} [mV]	U_{90} [mV]

Other parameters:

3.2 Repeat measurement of Bandwidth-length of interconnected reels and end-to-end setup. Modulate LED in the range of 2 MHz to 28 MHz. Obtain and draw by hand results for fibre frequency response

Setup: interconnected

f [MHz]	U_{out} [mV]

Other parameters:

Setup: end-to-end

f [MHz]	U_{out} [mV]

Other parameters:

Name: _____ Student Number: _____

HW-2. Eye Diagrams and BER

See Fig 2.3, Eq (13), Fig 2.6, 2.7,

Experiment 6.1 BER(COM) Manual pp 21
6.6 22
6.7 23
6.8 24
6.9 25
Results table 44

6.1 Set up pseudo-random bit-sequence PRBS generator at 10,20 and 40 Mbit/s
CRO at 2 V/div or 1 V/div (uncal) and 5 ns/div
Observe eye-diagram
Measure tr and tf at each bit-rate
With cursor on, or by visual inspection expect about 10 ns at 40 Mbit/s for example.

6.2 LED TX at 50 mA and 6 microwatt optical at patchcord
50 ohm CRO termination to avoid reflections (dips)
Measure tr, tf (manual p21) Expect about 11.5 ns or 45% at 40 Mbit/s

At 80% of v1measure t-pulse expecting about 16.5 ns or about 64% of bit-period

Measure jitter on eye expecting about 500 ps

6.3   Use template p28 for results
      Measure tr and t-pulse for 1,2 and 3 km cable
      Example: at 3 km + patchcord t-pulse of about 23.5 ns

      At 50% level measure full jitter; expect about 1ns after 1 km and 2
      ns after 3 km
      This due to amplitude noise converting to time-spread on sloping
      rise/fall

6.4   At 1,2 and 3 km measure eye Q factor or (v1-vt)/vN
      where vN is noise spread of v1 at 20 mV/div
      Expect about Q = 6 for 3 km

      Use Fig to estimate BER from Q^2
      Expect about 0.8×10^{-9}

      Measure jitter; expect about 2 ns due to amplitude noise converting
      to time spread
      (Was about $\frac{1}{2}$ ns at 6.2)

6.5   Repeat 6.2 to 6. 4 with laser TX at 33 mA, 10 to 40 Mbit/s, about
      520 microwatt

      Then insert 1 to 3 km cables and tabulate per template

      Expect tr about 10.5 ns; and 15.5 ns for 3 km

      Measure jitter (expect about 1.5 ns) and Q
      Then estimate BER   Expect Q about 7.5 giving BER 10^{-12}

      Tabulate results.